Housekeeping Management for Hotels and Residential Establishments

Uniform with this volume

Services and Maintenance for
Hotels and Residential Establishments

by Rosemary Hurst

Housekeeping Management For Hotels and Residential Establishments

Rosemary Hurst
M.H.C.I.M.A., A.M.B.I.M.

HEINEMANN : LONDON

William Heinemann Ltd
15 Queen St, Mayfair, London W1X 8BE

LONDON MELBOURNE TORONTO
JOHANNESBURG AUCKLAND

First published 1971
Reprinted 1975
434 90790 1

Printed in Great Britain by
Cox & Wyman Ltd, London, Fakenham and Reading

INTRODUCTION

The Hotel and Catering Industry is Britain's fourth largest industry and as such plays a large part in both the economic life of the country and in the encouragement of good international relationships through its contact with the ever-increasing number of tourists who stay in our hotels and holiday centres.

The industry is one in which there are very many different aspects, with members operating and managing in hotels, hospitals, schools, colleges, and universities, as well as providing a wide range of welfare and supporting services. Within the industry, housekeeping plays an essential part in providing clean, comfortable, and attractive premises and accommodation in such a way that guests and residents are made to feel welcome and to want to return.

This book, and its companion volume *Services and Maintenance for Hotels and Residential Establishments*, have been planned to cover all aspects of housekeeping and maintenance and to meet the requirements of the relevant parts of the syllabus of the Hotel, Catering, and Institutional Management Association. The books are also intended to provide a framework for all students entering the industry whether they wish to work in hotels, clubs or other residential establishments.

In an analysis of hotels listed in guide books, which the Hotel and Catering Industry Training Board quote in their approach to Front Office and Reception Training, nearly seventy-five per cent of hotels have fewer than forty bedrooms – only six per cent have more than one hundred bedrooms. In a great number of these smaller hotels, the responsibility for housekeeping and for minor maintenance is frequently undertaken by the wife of the manager or proprietor or by a competent assistant – often with no formal training. It is hoped that they will find the following chapters of use when they come to consider staffing questions, the control and costing of cleaning services, buying, and replacements, and the many problems which can occur in any housekeeping department.

Some years ago, on taking up an appointment as Resident Manageress, I was given a copy of the job description. I think it is of interest as it summarizes and lists the main duties, emphasizes what to a large firm, are the most important aspects of the job, and clearly states the limits of authority. It is as follows:

1. *Staff.* Control of recruitment, including the organization of duties and the discipline of the domestic staff – any increase in establishment must have prior approval. Control of the work of the pay-clerk/office assistants whose duties include the preparation of wages for the domestic staff, control of the cash books, purchase orders, and the student and visitor registers; operating the telephone switchboard and all general clerical duties.

2. *Domestic Expenditure.* (*a*) Directly responsible to the finance department for the imprest account which at present is fixed at £*xxx*. (*b*) Ensuring the operation of internal controls in the domestic expenditure. (*c*) The purchase and control of domestic goods and consumable stores with the authority to place orders up to £*xx* for any one item – such expenditure to come within approved revenue estimates or approved capital expenditure as provided by the budgetary control procedure.

3. *Buildings and their contents and the grounds.* (*a*) Responsible for the day-to-day condition of the building and the grounds. (*b*) Responsible for the furniture, fittings and fixtures, the carpets, cutlery, kitchen equipment, and other contents. (*c*) Ensuring that all contract work in the building is carried out as required. (*d*) Ensuring that all rooms are clean, comfortable, and in good condition. (*e*) Ensuring that the grounds are kept tidy.

4. *Meals.* The preparation of menus. Ensuring that meals are served on time and are of a satisfactory standard.

5. *The well-being of all visitors.* (*a*) Generally responsible for the health, care, and comfort of lecturers, guests and course members whilst in residence. (*b*) Being available to deal with any queries or difficulties they may have over meals or accommodation.

6. *Miscellaneous.* Any other duties that may be properly assigned to the Resident Manageress.

Housekeeping Management and *Services and Maintenance* cover most of the above points with the exception of meals and kitchen control and the up-keep of the grounds; these are complex subjects and need to be dealt with separately.

In both books, I have tried to explain the principles and ideas behind all planning and control.

ACKNOWLEDGMENTS

Without advice and encouragement from many friends this book would not have been completed; to them I am very grateful.

I wish to thank particularly the many firms who have given assistance and also the following who have given permission for the reproduction of diagrams and illustrations which I have needed:

The Architects' Journal (1.4)
Permission from the Controller of Her Majesty's Stationery Office for an illustration from:
Space in the Home (metric edition) 1970 (9.1)

R.H.

CONTENTS

1 FURNITURE

At a recent conference to discuss the provision of furniture for a student's hall of residence, one ratepayer was surprised to discover that part of the specification was that it had to be strong enough to withstand the onslaughts of a rugger team. Furniture for hotel and institutional use does not get the same gentle care that it does in the home; furniture produced for 'public' or *contract* use is designed to withstand harder wear and to retain a good appearance longer than the furniture made for the domestic market.

As there are more than 1500 furniture firms in this country and a considerable number of firms which import from overseas, there is a very wide range from which to choose; included in this number are the many specialist firms who make furniture particularly designed for hospitals, hotels, offices, and schools. A guarantee of reliability and good design is provided by those manufacturers who are approved by the Council of Industrial Design and by furniture which carries the 'kite' mark (*see* Figure 1.1).

FIGURE 1.1 The B.S.I. Kite mark and label of the Council of Industrial Design

GENERAL POINTS TO CONSIDER WHEN BUYING

When buying the main considerations are:

1. *DURABILITY*

(*a*) Construction should be strong and robust with firm well-fitted joints which do not 'give' if pressure is applied. The most usual joints are mortice and tenon, dowel, tongue and groove, and combed joint (*see* Figure 1.2).

These joints should be unobtrusive but strongly constructed so

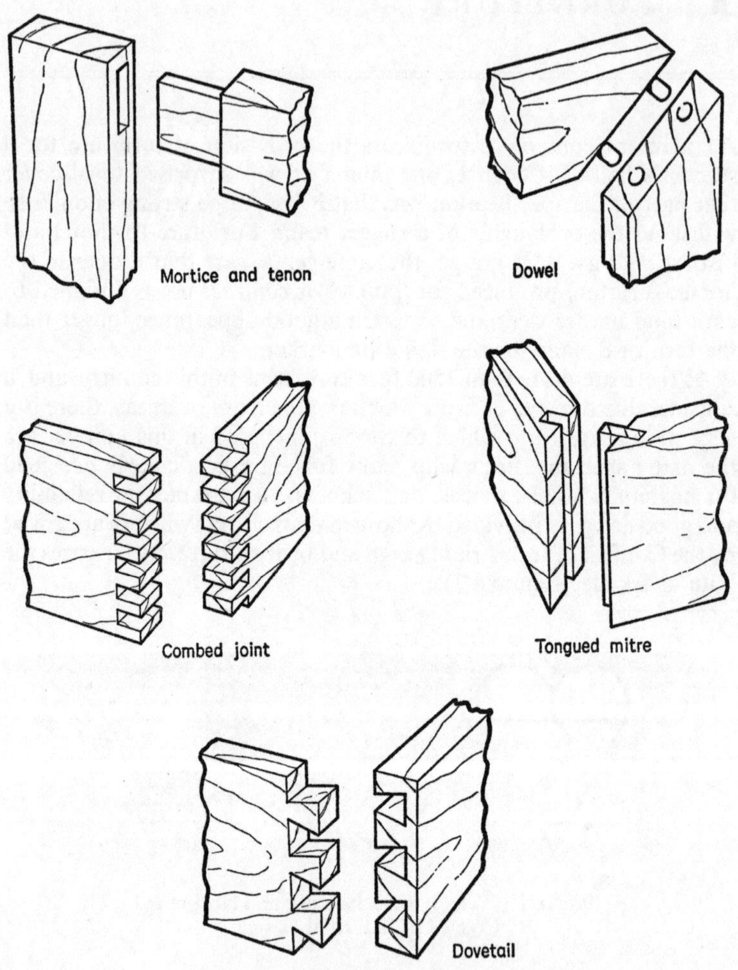

FIGURE 1.2 Different types of joints

that when buying there is the need to look at the back and underneath of furniture to find out how well it has been constructed and to take out drawers to find any weak points. Doors and drawers should fit and open smoothly without sticking, whilst all handles should be solid and firmly attached, hinges strong, and locks and keys turn easily. If buying in bulk, care should be taken to ensure that all locks are different and that duplicate or replacement keys are available.

(*b*) Surfaces should be stain, heat, and water resistant. The most usual treatment is a synthetic plastic lacquer, obtainable in matt or gloss finish, which is highly resistant to marking. Another alternative is the use of plastic sheeting, formica or melamine type; these are extremely difficult to damage being heat, scratch, acid, alkali, stain, and water resistant. Colours are varied and can imitate the appearance and texture of a polished wood surface. The vulnerable point for plastic sheeting is at the edge which can be knocked or chipped or can lift away from the base wood; it should be protected either by a wood edging or by careful fitting.

Plate glass is frequently used to protect the surface of hotel furniture; it can, however, be broken or become lost and, as dust can accumulate under it, it has to be moved for cleaning purposes.

(*c*) Finish of the furniture should be smooth with no rough surfaces, unevenness, or sharp corners which could cause damage.

(*d*) Upholstery fabrics should be strong, not of a type that is easy to catch and pull, and easy to clean either by shampooing or by dry cleaning with a cleaning solvent. Colours should be fast, and not likely to fade with strong sunlight.

2. *CONVENIENCE*

(*a*) Ease of cleaning. The design should be simple with no unnecessary edges, carvings, or angles in which dust and dirt can accumulate. Internal angles should be coved to ease cleaning.

(*b*) The height from the floor. The furniture should either be a minimum of 203–228 mm (8 to 9 in.) from the floor so that cleaning equipment will go under easily, or else the dust problem is solved by the use of fitted furniture.

(*c*) The weight of the furniture and its ease of movement. Are castors necessary? If so, are they adequate for the weight and are they big enough to run freely from one floor surface to another?

3. *DESIGN AND PROPORTIONS*

Is it suitable for the use for which it is intended, and for the type of establishment? Is it of a suitable size to match the proportions of the room? Small rooms need furniture which is also small, whilst a large impressive room will require furniture on the same scale.

4. *COMFORT*

Particularly for beds and chairs.

With the limited space that so many establishments have and their multi-purpose use, consideration must also be given to the choice between *built-in, fitted furniture* and *free-standing furniture. Built-in furniture* has many advantages:

(*a*) It can be designed for a particular purpose and so tailored that it takes advantage of recesses or alcoves or any extra inches which are often available in a room.
(*b*) It is often designed to hide and contain wash-handbasin units.
(*c*) By building to the ceiling, extra storage space is made available.
(*d*) It is easy to clean.
(*e*) The appearance of the room is stream-lined and neater.

The main disadvantage may be the initial cost but this is very dependent on the contractors and at what stage the furniture is fitted: if it is part of the original building plan and is completed with the contractor's building and finishing programme, it may well be cheaper than the cost of free-standing furniture.

To install built-in furniture in an older building may be more difficult as it has to be firmly attached to the walls; many walls in old buildings do not run true, ceiling heights can vary within a room, and corners are not always square. Another problem met in older buildings is that rooms are not always uniform in size or shape so that the furniture required may differ from room to room.

A further disadvantage is that built-in furniture is immovable and cannot be easily transferred from one room to another; this offers little scope for the individual rearrangement of a room.

In the *Hostel User Study* made by the Building Research Station, the discussion and comments on hostel accommodation 'showed that young people like to be able to move desk and bed around to take advantage of seasonal views, make best use of sunlight and of electric lighting arrangements, and have a change; and in particular, to make rooms different'.

Whilst most furniture is still traditionally made from wood, other materials are being increasingly used, mainly metals and the moulded plastics or polyester glass fibres. Wood itself is used either in the solid piece or as ply or laminated wood. These may or may not be covered by a more expensive wood veneer or by a protective plastic made to imitate the wood grain or as a contrast. Blockboard, battenboard, laminboarding, and some of the cheaper softwoods can also be used as a base for veneer or plastic.

Plywoods and blockboards are used to lighten the weight of furniture, often to give extra strength. Plywoods and laminated wood can be shaped and bent more easily than solid wood to fit a particular design.

CHOICE OF FURNITURE

CHAIRS

The most comfortable chairs are those designed to fit the body measurements of a particular user; this is manifestly impossible for hotels and institutions to provide as they must buy for the average user. The basic principles which govern the choice of chair are:

1. Dimensions

(*a*) The seat height related to the lower leg length. Anthropometric data tell us that for a man between 18 and 40 years this should be 435 mm (17 in.) and for a woman of the same age group 418 mm (16½ in.), which reduces to 404 mm (16 in.) for the 60 to 90 age groups.

(*b*) The seat depth related to the upper leg length. The average for a man is 478 mm (18¾ in.).

(*c*) The seat width related to the hip spread. For a man 366 mm (14½ in.), for a woman 391 mm (15½ in.). This would be adequate for a dining-room or upright chair but for comfort the width should increase to about 508 mm (20 in.) for an easy chair.

(*d*) The angle between the chair back and the seat should be related to the natural sitting position. For comfort and to give proper support, this is an angle of 105°; but an angle of 95° may be acceptable to some users.

(*e*) The chair seat and its angle should relate to the body weight and dimensions. The seat, on average, should be 7° from the horizontal. If it is horizontal, there is a tendency for the body to constantly slip forward so that frequent adjustment of the body muscles is necessary.

(*f*) The height, length, and position of the chair arms.

2. The balance and stability

This is improved by slightly splayed legs at the back of the chair, by non-slip rubber ferrules or caps fitted to the legs, and by arms not projecting more than 25mm (1 in.) beyond the front of the chair.

3. The height

The chair must be related to the height of the table, desk, or dressing-table with which it is to be used. Clearance between the seat top and the table should be about 254 mm (10 in.) to give minimum comfort for a man.

4. Comfort

All chairs should be used for one to two hours before a decision is made, particularly when they are for office or lecture-room use, as a chair which seemed reasonable when first used can develop many painful areas after one or two hours.

Easy Chairs

Cushioning is now of latex foam rather than the springs and horse-hair which were used for upholstery for many years. Latex is practical as it retains its shape, is not harmed by damp or upset liquids, and is moth and vermin proof. Cushioning should be slightly firm so that the weight is well supported.

Covering material should be hard-wearing, non-slippery, and easy to clean; darker patterned fabrics show the least reaction to dirt and fading. Covering and upholstery should be tested for flame retardency. Some upholstery materials cannot be wet cleaned and a solvent has to be used which may also spoil some fabrics. The most usual fabrics are tweeds, moquettes, most of the man-made fabrics, leather, and vinyl materials; but, whatever material is used, the upholstering should be tight with straight seams and well-secured corners. Seat supports must be strong and easy to replace. For wearing ability, the fabrics used by British Rail for their upholstery, moquettes and plush velvet, have been well-tested. In a chair, the main points of wear are on the arm-rests, head-rest, and the seat edge. To prevent some of this wear wooden or semi-wooden arms are practical. Note that when a dowel joint is used to join leg to arm (Figure 1.3), it can loosen in use and, if not repaired, will weaken the joint between the arm and the chair back.

Loose covers should be well-fitting with the seams following those of the original covers as a hard under-piping or roughness will quickly wear through the outer cover.

An easy chair should be very slightly flexible.

The minimum performance test for easy chairs is the BSI specification for upholstered furniture.

Upright or dining-room chairs

Upright chairs should be rigid and not creak or groan when tilted back. For heavy duty, the low-backed variety are more serviceable as those with a high back will weaken over the years and develop a wobble, as the greatest strain on a chair comes at the angle of the back and seat and on the back legs.

There are two British Standard specifications which are relevant when discussing this type of chair.

Dimensions recommended by BS 3079:1959 are: Seat height from floor 432 mm (17 in.). Seat depth 330 to 381 mm (13 to 15 in.). Back-rest, fixed: minimum height of lower edge 203 mm (8 in.), maximum height of upper edge 330 mm (13 in.), and the radius of the curve for the backrest 406 to 457 mm (16 to 18 in.). This differs from the BS 3893:1965 specification for office desks, tables, and seating which allows a seat depth of 355 to 470 mm (14 to 18 in.) and the radius curve for the backrest of 305 to 457 mm (12 to 18 in.). These figures are based on average body measurements. A chair which has similar

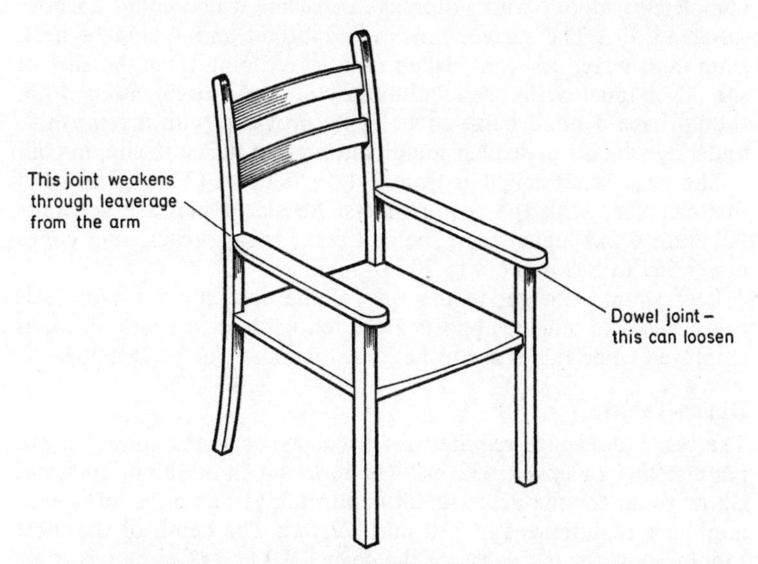

FIGURE 1.3 Dowel joint between leg and arm of chair

dimensions to these is more likely to be comfortable than one which differs markedly.

The appropriate test for strength is that of BS 3030, Part 11:1959 which relates to School Furniture.

As was mentioned earlier, the gap between the top of the chair seat and the table or desk with which it is to be used should be a minimum of 254 to 305 mm (10 to 12 in.); so it follows that, when buying, both pieces of furniture should be bought together or else careful measurements should be taken.

If chairs are moved frequently or have to be stacked, they should be light, very sturdy, and able to stack easily and quickly to give a stable pile which will not knock or topple over.

Removable seats make cleaning considerably easier and quicker but are not to be recommended if the chairs are moved often as the seats become detached, can become lost or damaged, and do not always fit back into other chairs so well.

When chairs are to be used for conferences or lectures it is well to consider those with a broad arm for note-taking use, with ash-tray attachments, or interlocking chairs to simplify room arrangement.

TABLES

Tables should be firm and well balanced. Tops should be level and completely smooth with no cracks or seams which could harbour dust and dirt. The surface finish is important and should be heat, stain, and water resistant; when used as a dining table the surface should be quiet when used without a cloth or, if used with a cloth, should have a fitted baize cover or be provided with a removable underlay which will deaden sound and prevent the cloth slipping off,

The most usual height is from 711 to 762 mm (28 to 30 in.) but this can vary with the requirements. Bedside tables are generally 101 mm (4 in.) higher than the mattress for easy reach, and coffee tables 355 to 508 mm (14 to 20 in.) high.

Legs should not get in the way of the user, and to avoid this cantilevered or single supports are often used; or a trestle fitting is employed when tables are to be joined together for banqueting.

Dining-Tables

The basic allowance required for each person is the space for the plate setting or cover, 432 mm (17 in.), and in addition, sufficient elbow room for the diner to eat comfortably; this gives an overall minimum requirement of 580 mm (23 in.). The depth of the table should allow for the width of the cover, 300 mm (12 in.), space for condiments and decoration, and also room for the diners to sit easily without their knees touching. This gives a minimum width for comfort of 750 mm (29½ in.).

Circular tables take more space than the conventional square or oblong tables but if a mixture of both types is used it will break up a large dining area to give a more individual effect. More people are able to crowd companionably around a circular table than can do so round one which is square or oblong; the allowance for a cover on this type of table is 609 mm (24 in.), this giving sufficient elbow room for the diners.

DRESSING-TABLES AND CHESTS OF DRAWERS
(*Storage space*) (*see* Figure 1.4)

Whether provision is made for both pieces of furniture depends on the length of stay of the residents. When residence is for the greater part of the year, good storage space is essential; whether it is best to provide this in the form of drawers or shelves is debatable but, where there is a constant change of over-night guests, it should be noted

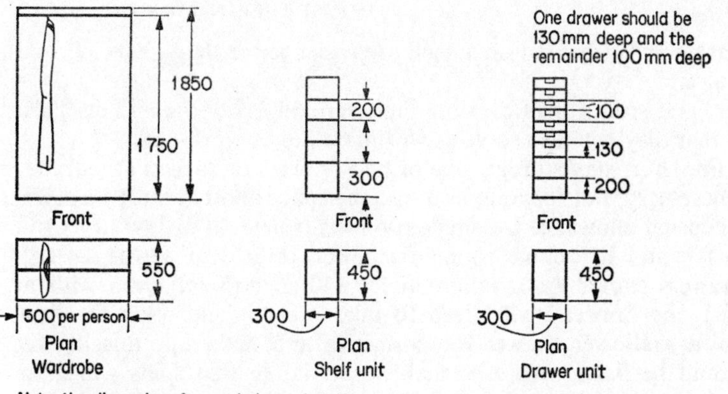

Note: the dimensions for wardrobe and shelf and drawer units are minimum internal clear dimensions

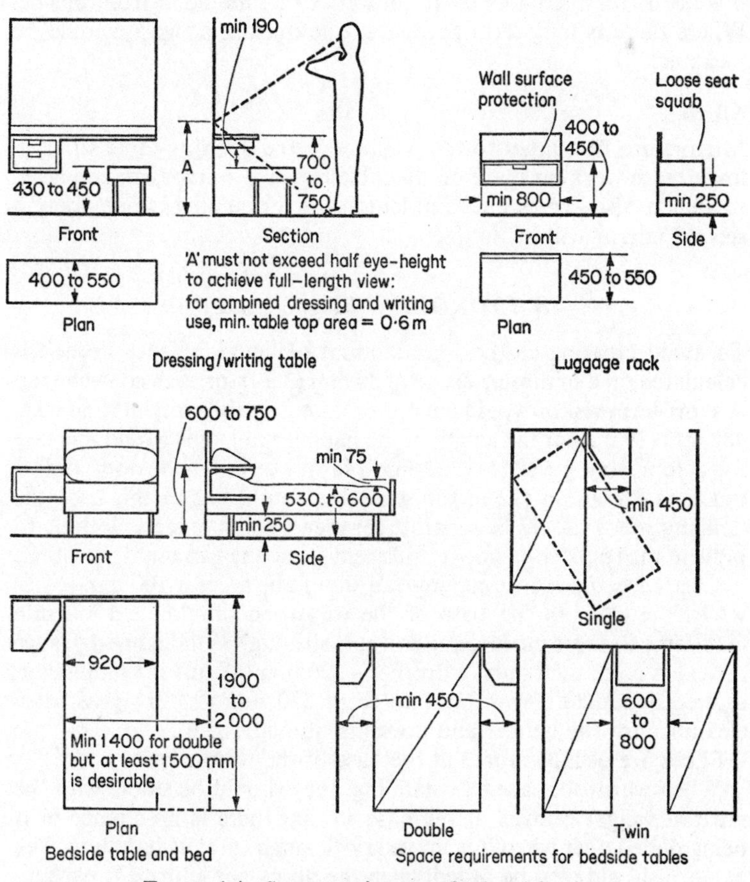

FIGURE 1.4 Space requirements for hotel furniture (From *AJ Metric Handbook*)

that items left behind on a shelf are easier to see than items left in a drawer.

A great number of dressing-tables are dual purpose and designed so that they will also serve as writing-tables or work-desks.

For short-stay storage, one or two drawers or shelves are all that is necessary, but for long-term use the space allotted must be more. The usual allowance for single rooms is from 1·30 to 1·49 m^2 (14 to 16 ft^2) and in double rooms from 1·86 to 2·23 m^2 (20 to 24 ft^2). Drawers should be a minimum of 130 mm (5 in.) deep with at least one drawer really deep to take sweaters and other woollens and a shallower drawer for cosmetics and make-up; this drawer should be lined with a washable material so that spills will cause no damage to the wood and cleaning is made easier. The modern practice is for the bases of *all* drawers to be moulded from plastic. Where there is long-term residence, one drawer, at least, should be lockable.

Mirrors

Mirrors are best fitted to the wall above the dressing-table so as to leave more working space on the table top. The mirror can be placed so that it also serves as a full-length mirror; if this is not done, a second mirror will be needed.

WARDROBES (hanging space)

To avoid creasing clothes, the amount of hanging space needed is calculated on a minimum space of 80 mm (3 in.) for each coat-hanger. A short-term visitor would have approximately six garments and need 480 mm (19 in.) for the length of the hanging-rail; this would increase for a long-stay guest to 1220 mm (48 in.) in a single room and as much as 1830 mm (72 in.) in a double room, but in this case the hanging space should be separate for men's and women's clothing to prevent the transfer of smoke and scent from one garment to another.

Clearance above the hanging-rail should be at least 100 mm (4 in.) whilst the drop to the base of the wardrobe should be 1750 mm (5 ft 7 in.) to meet most requirements; although if the hanging space provided is for men only, a drop of 1220 mm (48 in.) is usually considered adequate. From back to front 550 mm (22 in.) gives easy clearance for the hanger and prevents 'shoulder rub'.

Shoes are usually stored at the base of the wardrobe.

When wardrobes are free-standing, they should be stable with the greatest weight centred at the base so that there is no chance of it being pulled over when it is moved or through careless handling. The balance should also be tested when the doors are left open, particularly if one door carries a mirror fitting as this, plus the weight of clothes draped over the door, could make the wardrobe top-heavy; it also places undue strain on the hinges.

BEDS AND DIVANS

As bedrooms are increasingly used as sitting-rooms, temporary meeting places, and offices, the bed comes in for hard use as it doubles for seating or as an extra table; but, since it must still remain comfortable for sleeping, quality and durability are all-important.

Standard sizes are 150 × 200 cm and 135 × 200 cm (59 × 79 in. and 53 × 79 in.) for a double bed, and 80 × 192 cm and 100 × 200 cm (32 × 79 and 40 × 79in.) for a single bed. In hotels, double beds are usually replaced by twin beds which can be locked together when needed. This gives more flexibility and provides for the individual taste for a soft or hard base or mattress.

The most comfortable height for getting into the bed and for sitting is about 560 mm (22 in.) to the top of the mattress, with a clearance from the floor of about 250 mm (10 in.) to make cleaning easier.

Castors must be fitted so that the bed moves easily for bed-making, but they should be chosen with some regard for the floor surface as one does not want a bed which rolls smartly across the room as soon as it is touched. A solution is either to have castors with a brake fitting or else to fit only one pair of castors, to the head of the bed with the leg length adjusted accordingly, so that the maid can lift the bed at the foot end and move it easily.

Beds

A bed consists of the bedhead and foot end joined together by two metal rails with a mattress support of wire mesh or open spiral springs. This support is level and firm and covered by a hessian or felt underlay so that the mattress is protected from friction or rubbing. Both types of support are dust collectors, and need careful periodic cleaning with a suction cleaner.

These beds are easy to take down and store.

Divans

Divans have an upholstered base which is interior sprung; there may or may not be head and end boards. Legs are either built in as a part of the base or screwed in; in either case they must be firm and rigid.

Bed bases should be firm to give good support to the sleeper; following the trend for a firmer base is the use of laminated-wood strips or of removable wooden boards slotted into the bed frame, although these will have more use in a college or university residence than for hotel use.

Bedheads are a necessity to keep the pillow in position and must be large enough to protect the wall behind the sleeper so that it does not become discoloured with grease and dirt. If the bedhead is detached and fitted to the wall, there must be clips or other attachment to hold the bed firmly in place. Headboards must be easily cleaned.

End boards, if used, must allow for a tidy finish when the bed is made.

Bed-boards

Bed-boards should always be made available for those people who need a particularly firm mattress base; these boards are placed under the mattress.

MATTRESSES

1. Interior-sprung

Interior-sprung mattresses (*see* Figure 1.5) have springs which are either 'open coil' or 'pocket springs'.

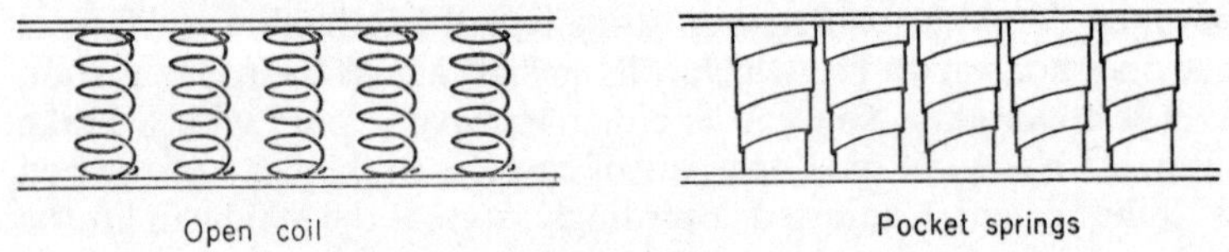

FIGURE 1.5 Interior-sprung mattresses

Where there is heavy use, pocket springs give the best service. The British Standard is 1877, Part 3:1963 and gives the minimum number of springs and content. With open-type springs the minimum is 192 springs for a single and 288 springs for a double mattress although a double mattress may have as many as 1000 pocket springs. Different grades of spring give extra support where it is needed; edges should be reinforced by a wire 'frame' so that the springs are protected if the bed is used for sitting.

Surrounding the springs are layers of cotton waste, curled hair, or polyurethene foam and, in the more expensive mattresses, a top layer of fleece wool. Covering this is a strong ticking of cotton, rayon, or linen. The ticking is tufted or buttoned to secure the padding in place and prevent it moving.

Ventilation holes are let into the sides of the mattress which should have strong fabric handles to facilitate turning, as this type of mattress should be well aired at least once a week and turned, but it should never be left doubled as this will eventually displace the springs.

Interior-sprung mattresses have a well deserved reputation for comfort but have the disadvantages of weight, of being open to attack from moths and other vermin, and of being absorbent; water from a burst water-bottle may take several days to dry out completely.

2. Latex or Polyurethene

These are a cheaper mattress made from either natural or synthetic rubber or from polyurethene.

The substances are foamed, poured into mattress moulds, and vulcanized. These mattresses vary in quality for firmness and thickness but are generally softer than an interior sprung mattress, lighter, and easier to handle. Their great advantage is that they are self-ventilating and require no turning as air is forced through the foam cells every time the bed is used. There is normally a right and wrong side (*see* Figure 1.6).

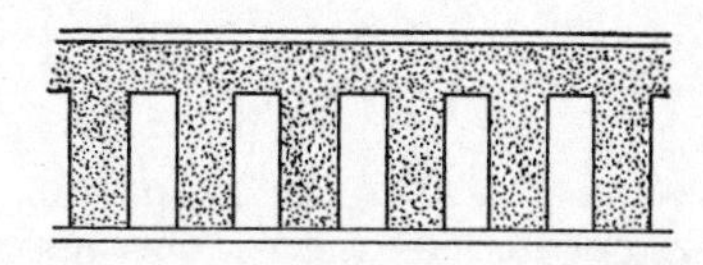

FIGURE 1.6 A latex-moulded mattress

These matresses are moth and vermin proof, make no dust, and clean and dry easily.

Because these mattresses are softer, it is recommended that they are used with a firmer base than would be necessary for a sprung-interior mattress.

3. Stuffed

Stuffed mattresses of hair or flock are still in use but are now rarely bought as they are heavy, can become lumpy, and are very absorbent and liable to moth and vermin attack. They should be remade every 4 to 5 years.

2 SOFT FURNISHINGS

LINEN-ROOM staff should be competent to make the simpler soft furnishings that are necessary; they should be able to measure, cut, and make curtains and the easier loose covers and bed-spreads as they are needed from time to time. Many suppliers now offer a free making-up service; this is an obvious saving provided it is not in exchange for a loss of competitive price, quality, or discount rates.

CURTAINS

Good window dressing can alter the proportions of a badly designed window by either making the window appear to increase or decrease in size or by drawing two or more windows together (*see* Figure 2.1).

Curtains are used for a variety of reasons, the most important being the provision of privacy, but others of almost equal weight are the prevention of heat loss and the exclusion of draughts, the deadening of outside noise, the need to hide an unattractive view, the reduction of direct sunlight in the room to prevent glare and excess solar heat gain, and the protection of carpets and other furnishings from fading.

Curtains are useful as room dividers and are an easy quick way of blocking off part of a hall, lounge, or restaurant, or as a covering to a wall for decorative effect. The material chosen must be suitable to the architectural style of the room, with the pattern in proportion to its size. Colour is chosen to blend or contrast with the decorations.

Fabrics suited to large important rooms are brocades, satins, tweeds, velvets, and corduroys, and for smaller rooms or study-bedrooms there are the chintzes and linens, repps, casement cloth, and folk weaves.

Bathrooms require an absorbent washable material, such as towelling, or else spongeable material, such as plastic or glass fibre. Some plastic materials stiffen and tear easily with use; glass fibre material can also cause a problem as it can crease badly when laundered and will rub and wear easily if hung against a projection.

An establishment may have different weights of curtaining for winter and summer use. A large establishment will have duplicate pairs of net curtaining so that they can be replaced when laundering is necessary. All net curtains must be laundered at the same time to avoid an obvious 'clean, clean, dirty' effect from the outside.

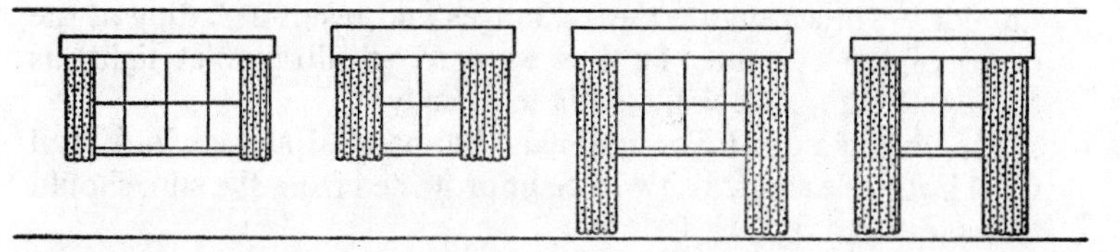

Window proportions can be increased or decreased or the windows joined together by the different uses of curtaining–

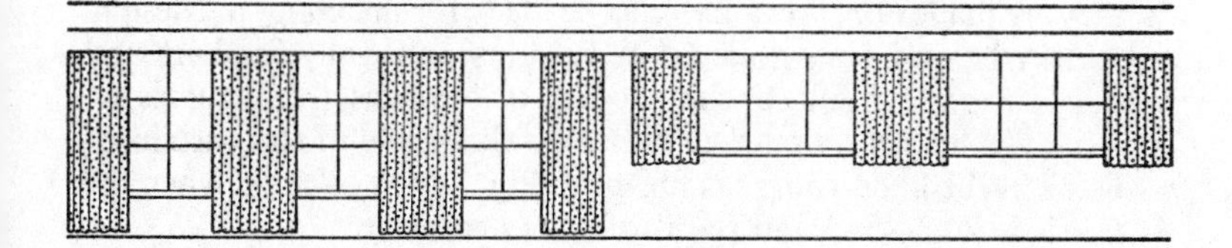

–or the curtains can be used as a room divider or to cover an entire wall

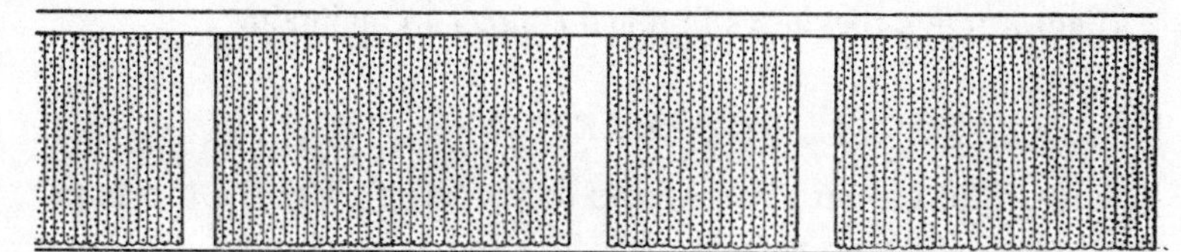

FIGURE 2.1 Curtains and window dressing

GENERAL POINTS TO CONSIDER WHEN BUYING CURTAINS

1. Always have the material hung horizontally so that the effect can be seen as it will be in the room for a pattern can look quite different when hanging in folds. The material should be seen in both natural and artificial light.
2. An allowance may have to be made for shrinkage. With material with a shrink-proof finish, shrinkage should be nil up to 1%; otherwise, it may be anything up to 5% or 6%, an amount that the Retail Trading Standards Association considers to be reasonable. Shrinkage depends on the looseness of the weave, on the amount the material has been stretched during manufacture, and on the washing conditions. If there is any doubt, it is as well to test the material.

3. Colour fastness to light, laundering, and cleaning, should be considered. Under certain conditions most colours will eventually fade, particularly in areas where there is high air pollution, sea air, gas fumes, or smoke. Sun (and light) with any condensation or humidity will accelerate these changes and result in fading at the edges of the curtains. Fabrics exposed to ultra-violet light, as through 'vita' glass, will also fade quickly.
4. If the curtains are to be unlined, the material should look well from both sides and, as it will be unprotected from the sun, should be guaranteed against fading.
5. Pattern should be parallel to the weft of the material; this is not always so and can lead to great difficulty when making up. As the pattern must be uniformly in line across the curtains when they are made up, allowance must be made for matching the design.
6. It may be company policy that curtains, especially for bars and public rooms, should be fire-proofed or fire-resistant. If this is so, it should be clearly marked either on the inside of the curtain or listed in the linen-room, as fire-proofing may be affected when the fabric is laundered and re-proofing is necessary.
7. The ease of laundering. This can be an expensive item, especially if dry cleaning is required. Some materials are drip-dry/non-iron; although it is often found that they look very much better after light ironing, they are still much simpler to maintain.

MEASUREMENTS

The length of the curtain is known as the 'drop', with the measurement taken from the curtain track or rod.

Curtains are made either

(*a*) *Full length.* These are finished 13 to 25 mm (½ to 1 in.) from the floor to prevent dirt and friction wear.
(*b*) *To the window-sill.*
(*c*) To 152 mm (6 in.) *below sill level* when the sill is more than 457 to 610 mm (1½ to 2 ft) from floor level.
(*d*) When curtains are hung *in the reveal* of the window, the finish is 12 mm (½ in.) above the sill.

The width of the curtain is called the 'head'; the minimum allowance for the head or fullness is 1½ times the width, not of the window itself, but of the track or curtain rod. For turnings, an allowance of 152 mm (6 in.) is added to the drop measurement but this may be doubled where it is suspected that the fabric may shrink, and the extra material is made into a double hem which can be let down when required. Extra material must also be allowed so that a pattern can be matched across the pairs of curtains.

For example: A window 2·286 m × 2·743 m (7½ ft × 9 ft) would need three lengths of material 1·219 m (48 in.) wide, each 2·296 m (9½ ft) long, requiring a total length of 6·9 m to make a pair of curtains each of 1½ widths.

LININGS

It is only necessary to allow for the drop measurement, with no extra allowance for turnings.

NET CURTAINS

As these are so much lighter, double fullness is allowed and 305 mm (12 in.) added to the drop. This gives enough material for a hem slotted on to rods or wires at top and bottom to prevent the curtain blowing about, or else for a double or treble hem at the bottom to give the extra weight needed for it to hang well.

Net curtaining can also be obtained with a metallic backing. This is aluminium particles bonded to the polyester fabric. Depending on the weave, the material reflects from about 45% to 65% of the sun's radiation so that solar heat gains are greatly reduced. In winter, the situation is reversed as the curtain reflects internal heating back into the room and also lowers the amount of heat lost through the window. These curtains are being increasingly used in modern buildings which are liable to excessive heat gains. The manufacturer's instructions for laundering this type of material should be carefully followed.

MAKING-UP

All good quality curtains are hand sewn as this gives a very much better finish. It is particularly necessary for velours and pile fabrics, brocades, and silks as the stitching remains unseen; machine-stitching on this type of fabric looks unsightly and is also apt to 'pull' the material.

In large establishments where expense and time have to be considered, there is often a compromise and the hem will be hand-sewn whilst the rest of the curtain will be machine-made. For the best results, however, both hems and edges are hand-sewn.

Expensive and good quality material is always lined as the curtain will hang and drape better and some protection is given to the fabric. The lining is usually of sateen which can be beige or cream or to match the curtaining. For heavier curtains an interlining may also be used; this can be either a flannelette material, 'bump', or else a metallic fabric, milium. This is a high quality cotton sateen with the reverse side coated with tiny aluminium particles which reflect the heat back into the room and act as an insulator.

UNLINED AND NET CURTAINS

1. The material is measured through, checked for the correct measurement, and cut square to the material, by following the weft thread. Care must be taken that the pattern matches across the curtain and that the head of each curtain is marked as it is cut, because from the head of the material the run must all be the same way; with many fabrics it is strikingly evident if this is not done.
2. Widths are joined with a run-and-fell seam (*see* Figure 2.2).

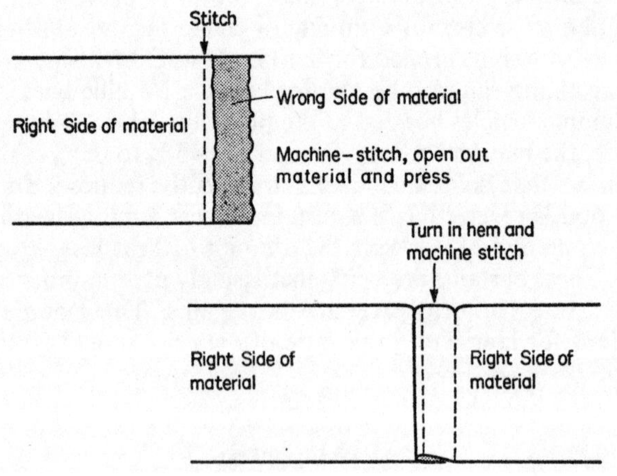

FIGURE 2.2 To join material using a run-and-fell seam

3. Turn under side seams: 13 to 25 mm (½ to 1 in.) for net curtains, 25 to 38 mm (1 to 1½ in.) for heavier curtains. To prevent puckering, the selvage should be nicked before turning (*see* Figure 2.3).
4. Lay material flat, press in bottom hem 50 mm (2 in.) wide, mitre corners, tack and hand or machine sew. Always machine net.
5. Measure upwards from the hem for the required drop measurement and turn in 25 to 50 mm (1 to 2 in.) (for the heading. This is dependent on type of heading and rufflette tape used (*see* below). Press and tack. Mitre corners.
6. The 'gather' or rufflette tape is machined along this hem taking care not to work over the strings of the tape. On the inside, or centre edge of the pair of curtains, these strings are sewn down firmly but they are left free on the outer edge so that, when finished,

the curtains can be gathered or drawn up to the window width and the ends tied firmly. These strings are released for laundering so that the curtains can be ironed flat.

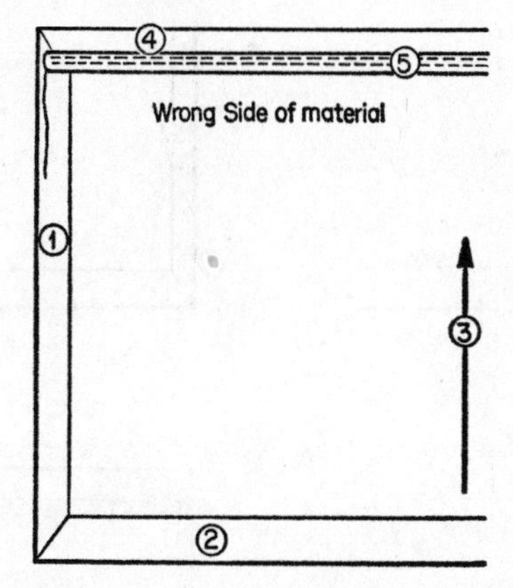

FIGURE 2.3 To make unlined curtains

1. Nick selvage edge, turn in, press, and hand or machine-stitch.
2. Turn up bottom hem, press, and tack–mitre corners–hand-sew.
3. Measure up from hem for required drop measurement.
4. Turn in, press, and tack in position.
5. Tack rufflette tape in position and to cover the raw edge of turning. Machine-stitch–leaving gathering threads free so that they can be pulled to gather the curtain to fit the window.

LINED CURTAINS (*see* Figure 2.4)

Measurements are the same as for unlined curtains but the lining is cut with no hem or turning allowance.

1. The bottom hems of the curtain and lining are pressed and hand or machine-sewn. This hem may be up to 152 mm (6 in.) deep depending on the size of the curtains and the weight of the fabric.
2. The side hems are pressed in to a depth of 25 to 50 mm (1 to 2 in.). With the selvage nicked to prevent puckering, the curtains are laid flat on a table and the lining is pinned and tacked into position, working from the bottom, with the lining hem 25 to 50 mm above the edge of the curtain hem. Mitre corners. Hand or machine-sew.

   With extra large curtains taking 1½ or more widths of material, the lining *must* be tacked in position at a 300 mm (12 in.) spacing; otherwise it is extremely difficult to get the lining straight in relation to the outer material, if this is not done the lining will sag or billow.
3. The required drop is measured upwards from the bottom hem and the process continues as for an unlined curtain.

With lined curtains, small weights are often sewn into the bottom corners or a length of chain is inserted in the hem to assist the drape of the curtain and to keep it in place.

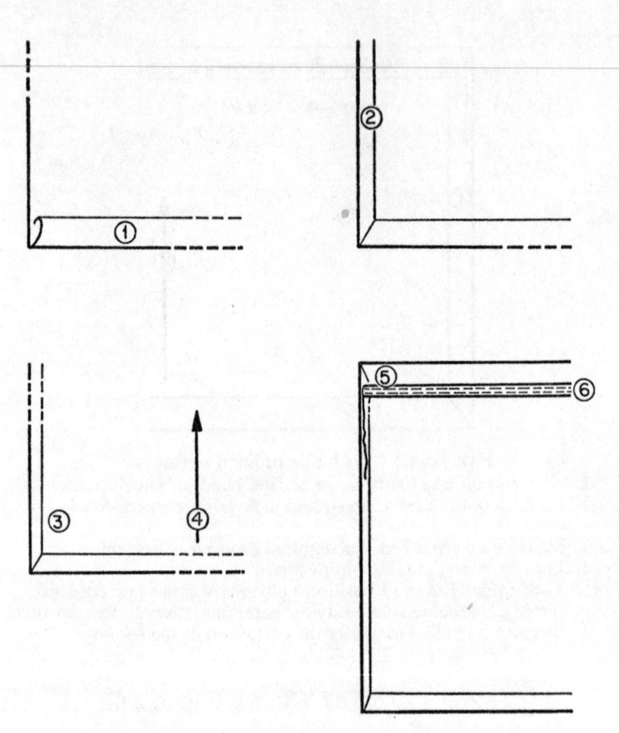

FIGURE 2.4 To make lined curtains

1. Turn up bottom hem, press, tack, and hand-sew. Repeat for the lining–but this hem can be machine-sewn.
2. Press and tack side hems in position–mitre corners.
3. Tack lining in position working from bottom hem upwards with lining hem 25 to 50 mm below edge of curtain hem. Hand or machine-sew.
4. Measure up from hem for required drop measurement.
5. Turn over, press, and tack top hem in position. Mitre corner.
6. Tack rufflette tape in position taking care that the raw edges of both curtain and lining are covered. Machine-stitch leaving gathering threads free so that the curtain can be gathered to fit the window.

Detached Linings

Linings are not always sewn in as one with the curtain but can be made separately and attached by means of hooks and a curtain lining tape. Curtains are made in this way if there are laundering problems with the outer material or if the curtains and linings shrink at different rates. They may also be made in this way if the curtains are several widths wide as these may be very heavy to handle and difficult to adjust for a sewn-in lining.

The same lining can also be used for both winter and summer weight curtains.

TYPES OF HEADINGS

Tape gathers the curtain smoothly to the curtain track. Different types can be obtained to pleat and stiffen the top of the curtains so that the curtains can be hung without a pelmet with the heading sufficiently deep to hide the track.

For many of these different tapes, double or treble the width of the curtain track must be added to the measurement of the curtain material.

CURTAIN TRACKS AND RODS

Originally, curtains were hung from a cornice pole often 75 to 100 mm (3 to 4 in.) in diameter. The modern track is much lighter and is fitted to the wall or ceiling. Many tracks have concealed fittings and, as they match the paint-work, are almost invisible. They should be fitted so that the curtains hang within 50 mm (2 in.) of the wall.

Pelmets, when used, are of wood and usually painted to match the decorations. If the top is boxed in, this helps to keep the curtains dust-free.

Pull Cords are easy to fit and prevent dirt accumulating on the curtains through constant handling.

OTHER WINDOW COVERS

VENETIAN BLINDS

Their main uses are to reduce and control excessive sunlight and heat entering a room, to protect the carpets and furnishings, and to give privacy. They are made from light alloys or from coloured plastics, the slats being horizontal or vertical. Venetian Blinds can be fitted externally or internally to a window.

Weekly cleaning is done with a soft brush; more thorough cleaning is frequently done under contract when the blinds are taken down and sent to a specialist firm; if the cleaning is done *in situ*, the slats are either wiped over with a damp cloth or else a specially designed cleaning tool is used which cleans both surfaces of the slat in one operation.

Maintenance

The cords should last about 4 years and the tapes about 7 years but this depends very much on the person operating the blind; some may be in perfect order after 20 years.

ROLLER BLINDS

These are becoming increasingly used as the choice of fabric and colour widens.

Clean by brushing thoroughly.

LOOSE COVERS

The materials used are similar to or the same as those used as curtaining, such as tweeds, repps, moquettes, and linens, but a further requirement is that the fabric should also be crease-resisting. The material should be seen flat in position on the furniture before deciding if a pattern is suitable or not.

If there is any risk of the material shrinking, it should be laundered before it is cut – and care must be taken that the pattern is cut the same way of the material; this applies to both plain and patterned material, so that if any subsequent shrinkage does take place, the cover will not pull out of shape or tear. When there is a definite pattern, this should be centred or cut in the same place in all the covers being made to achieve uniformity.

The great advantage of loose covers is that they preserve the appearance of the furniture, make cleaning and laundering easier, and will make a collection of chairs and settees into a uniform set of furniture.

The greatest wear comes on the arms; to avoid this arm caps are often used. Wear is also caused by friction between the under-upholstery, particularly with stiff piping or cords and seams which rub against the loose covers; care must therefore be taken with the fitting.

Loose covers are expensive and should be kept well-laundered to prevent ingrained dirt and grease rubbing or cutting the material fibres. To reduce the time needed for refitting covers after laundering, each chair and cover should be marked, so that it is plainly evident where each goes. For fasteners, zip fastenings and hooks and eyes are effective but they may be damaged; tapes which tie under the furniture should be avoided as they constantly come off during laundering.

Sometimes, it may be cheaper to have furniture re-upholstered with a material which wipes clean, rather than go to the expense of buying new covers. It can also occasionally, be cheaper to buy a new piece of furniture.

As a rough guide, the following is the average amount of material needed for:

Easy chair	– 5·5 m (6 yd) of 1·2/1·3 m (48/50 in.) material – in a plain-to-floor style
	– 6·5 m (7¼ yd) – if box-pleated or frilled
Settee	– 9·2 m (10 yd) – plain-to-floor style
	– 11·0 m (12 yd) – box-pleated or frilled
Seat, box cushion	– 1·0 m (1 yd)
Back cushion	– 0·75 m (¾ yd)
Divan cover	–
(small) 100 × 200 cm	– 7·5 m (8 yd) – plain-to-floor style
(large) 150 × 200 cm	– 8·25 m (9 yd) - plain-to-floor style

An extra allowance is needed when a patterned material is used

Stretch covers, made from nylon and man-made fibres, are in general domestic use but are not sufficiently hard-wearing for hotels or other residential establishments; although the fabric is colour fast and requires no ironing, the material pulls easily and the surface rubs to give a shabby appearance in a relatively short time.

If loose covers are laundered on the premises, the longest and hardest part of the job, in many cases, is the ironing that is required. With many fabrics, if the cover is replaced on the chair when it is slightly damp, the ironing is then completed much more quickly with the chair used as a base; and frequently the result is better.

3 CUTLERY, TABLEWARE, CHINA, AND GLASS

CUTLERY

TRADITIONALLY, silver-plated cutlery has always been used where very good standards of service are maintained; but in the last twenty years or so good quality and well-designed stainless steel has been produced which can look equally attractive on a well-laid table. Silver plate is still bought for hotels, senior staff-rooms, and executive establishments where traditional beauty has its place along with white starched tablecloths or polished tables, fine porcelain, and sparkling glass. Stainless steel is found in most canteens, students' residences, and hostels, restaurants, and, when well-designed, in modern hotels.

SILVER

Silver on its own is a soft metal which is hardened by the addition of copper. Sterling silver is an alloy containing 92·5% silver and with the remaining percentage of copper; this carries a hallmark to show that it has been tested for quality at one of the four assay offices in London, Sheffield, Birmingham, or Edinburgh. Britannia silver has a higher percentage of silver of 95·84%. Both these metals are expensive and are rarely found in use in hotels or other large-scale establishments.

In general use is *Silver-plate*, or E.P.N.S. as it is commonly called, the letters being the abbreviation for Electro Plated Nickel Silver.

When utensils are silver-plated, by means of electrolysis, silver salts are deposited on blanks or bodies made of an alloy of either nickel-silver or nickel-brass; the quality of the resultant E.P.N.S. utensil depends on the thickness and strength of the base alloy, on the thickness of the silver deposit, and on the evenness of its application.

It is difficult to estimate the 'life' of E.P.N.S. as use varies so enormously from one establishment to another, but for hard wear a heavy deposit of silver is necessary; this is measured in pennyweight (dwt), which is 1/20 oz troy, for each dozen items of cutlery or for single items of tableware. Good quality plate will have these weights stamped on the back alongside the E.P.N.S. mark; tablespoons or forks are about 30 dwt, desertspoons and forks are 22 dwt, and teaspoons vary from 6 to 10 dwt depending on size.

E.P.N.S. can be repaired and when worn down to its base metal can be re-silvered; this is, however, expensive and would only be done if it were not possible to replace an original design or if re-silvering were cheaper than the replacement.

STAINLESS STEEL

Stainless Steel, like E.P.N.S., is available in different qualities and is an alloy of chromium, steel, and nickel. The British Standards specification for stainless steel gives the percentage of chrome and nickel in the steel to be between 18/8 to 18/10, which is 18% chrome to 8% or 10% nickel. 18/8 is considered the best quality to buy and is often specified in contracts both for cutlery and for kitchen equipment. Some cheap grades of cutlery are obtainable with 17% chrome (17/8), a chrome iron steel, but this has not the same resistance to corrosion as the better quality and does not retain its appearance so well.

Stainlessness is a relative quality as, in many cases, the steel will darken with excess heat, and so is not always suitable for use under a salamander or in a hot plate, but this stain can be removed with methylated spirit or occasionally a fine abrasive. Stainless steel can also discolour with salt/vinegar mixtures, chlorine-type bleaches, and some proprietary brands of 'silver-dip' if the articles are left in contact for any length of time. Staining can also occur when it is machine-washed; in time the cutlery is left with a dull lifeless finish which can only be removed by application of a special stainless steel polish. If cutlery is towel-dried and not left to air-dry after machine-washing this dullness may not develop, but it means extra manual work and some loss of hygiene standards.

The main points to be considered when buying stainless steel-ware are:

1. It is practically indestructible but, unlike E.P.N.S., when damaged it is very difficult to repair.
2. It is much lighter than E.P.N.S. which is of some advantage with waitress service.
3. No polishing is required, or only exceptionally.
4. Cost. The more expensive range of stainless steel overlaps with the cheaper ranges of E.P.N.S., but generally stainless steel is considerably cheaper.

Most manufacturers produce two finishes for stainless steel, a mirror and a satin finish; unless the mirror finish is specially protected (by a manufacturing process) it invariably turns into a satin finish after a period of time with wear and the innumerable small scratches on the surface caused by constant handling.

Design is very much a matter of personal choice; but all cutlery should be in one piece, avoiding separate handles which loosen or puncture so that grease and water can be trapped inside. Bone, nylon, or plastic handles can be affected by detergents or the high temperatures needed for machine-washing. Ornate design should be avoided as this prevents easy cleaning. To simplify sorting and selection, each item should be easily identifiable.

KNIVES

Main types are table, steak, fish, cheese, butter, cheese serving. They should be well balanced and comfortable to use with the handle fitting smoothly into the palm of the hand. The blade must keep a good cutting edge but a long blade is not necessary as the balance is in the handle. When it is laid the blade should not touch the table.

Fish knives have a broader blade than the other types and are not so sharp.

FORKS

Main types are table, dessert, fish, fruit, pastry. Modern forks can be either three or four-pronged with the three-pronged forks easier to keep clean but the four-pronged fork simpler to use. Bevelled edges to the prongs also makes cleaning easier. The strength of each prong is also important as this is where the damage occurs when machine-washed.

Hotels require for silver service that up to ten sets of knives and forks can be cleared at any one time in such a way that they will not fall off the plate. To do this correctly the knife blades are placed under the handles of the forks or the prongs of the fork under the spoon handles. In modern design there is an increasing tendency for this arch on the fork and spoon handles to be flattened, thus making clearing difficult.

SPOONS

Main types are table, dessert, soup, tea, coffee, mustard, and sauce and soup ladles.

Bowls should not be too shallow. Sizes should be in proportion with the other cutlery; tea and coffee spoons should fit well into the saucers and not be too long.

STOCK

There is an increasing tendency to rationalize all cutlery and reduce the number of different types in use and kept in stock; frequently, cheese knives and forks will be used for fish or the same spoon will double for tea and coffee. The degree to which this is done depends on the standards of service offered.

Cutlery losses can be staggering and careful checks should be made to prevent borrowing or other misappropriation and loss through careless plate clearing.

It is usual to have an arrangement with the swill buyer so that all teaspoons and other cutlery which has been inadvertently added to the pig's daily diet are returned.

In a restaurant or dining-room a minimum of 20% beyond the normal cover requirement is usually allowed but this may have to be substantially increased in a busy restaurant or canteen.

Store-room stock may be 40 to 50% of the amount in use for most cutlery but may well be 100 to 200% for teaspoons. Minimum stock figures will be related to the annual losses over the previous two or three years.

Where extra silver is required for occasional functions and special use, this can be hired if it is considered that the extra capital outlay is prohibitive.

BADGING OF CUTLERY

Badging of Cutlery is often the only positive way of identifying one's own property and is done at little extra cost by the supplier.

TABLEWARE

VEGETABLE DISHES – FLATS AND COVERS

These are generally of either E.P.N.S. or stainless steel as they are more durable than their china counterparts. Rectangular shapes are preferable to oval shapes as they stack easier and are better for arranging food portions. Covers should be so designed that they can also be used as a serving dish.

TEA SERVICES

These again are usually bought in E.P.N.S. or stainless steel to avoid the replacement of breakages for china, but it must be remembered that, although hardwearing, they can still be damaged through misuse. The most common damage is caused by lids coming off, or dents, and leaks from the jointing of the spout to the body of the pot.

Below are some standard sizes which have been found satisfactory from an economy/service ratio.

	To serve	*Size*
Teapots and Jugs	1 or 2 people	570 ml (1 pt)
	3 people	855 ml (1½ pt)
	4 or 5 people	1·14 litres (2 pt)
Cream jugs	1 or 2 people	171 to 199 ml (6 or 7 oz)
	3 or 4 people	285 or 313 ml (10 or 11 oz)
Sugar basins	1 or 2 people	199 or 228 ml (7 or 8 oz)
	3 or 4 people	285 to 399 ml (10 to 14 oz)
Vegetable dishes	1 person	180 mm (7 in.)
(round divided)	2 people	200 mm (8 in.)
	3 people	230 mm (9 in.)
	4 people	254 mm (10 in.)
Soup tureens	1 person	285 to 342 ml (10 to 12 oz)
	2 people	570 ml (1 pt)
	6 people	2 litres (3½ pt)
	12 people	4 litres (7 pt)

CHINA

This is a very general term for the articles which are used on the table and which are moulded or cast from clay materials and then fired.

The quality and grades of china, whether earthenware or fine bone china, differ from each other in the proportions of their ingredients and in the length of time and the temperature used in the firing.

EARTHENWARE

Earthenware is made from 25% ball clay, 25% china clay, 15% china stone, and 35% flint.

VITRIFIED WARE

Vitrified ware is similar to earthenware but has a greater proportion of stone. It is fired to a higher temperature with a longer firing time so that the particles fuse together more completely. The result is a much stronger material which will stand rougher usage. Naturally enough, it is a little more expensive.

Until recently, when disposable plastic and paper cups and plates became readily available, earthenware was used extensively for such outdoor catering as race meetings and annual shows because of its cheapness and the expectation of loss and damage; it is not really strong enough for constant canteen or cafeteria use.

Vitrified ware must not be confused with the *Vit* which can be found stamped on the back of some china. It can mean vitrified but can also mean vitreous which merely implies that it has been through a kiln. Price is a good indication of quality.

BONE CHINA

Bone china is made from 25% china clay, 25% china stone, and 50% animal bone. It is the bone which gives to bone china its great strength, its fine warm white look, and its translucency.

Articles are moulded or cast and left to dry. Later, when most of the moisture has evaporated and it is leather hard, the article is placed in the kiln and is *biscuit fired.*

The decoration can be added as either *on-glaze* or *under-glaze.* An *under-glaze* decoration is applied directly on to the article after it has been biscuit fired. It is then covered with a glaze and fired for the second time – the *glost* firing. The glaze, which is the bright shining surface on the china, is simply a glass which is annealed to the body of the article and completely covers it. The decoration shows through the glaze but the range of colour is limited owing to the high temperatures which are needed for the glost firing; these temperatures can alter or burn off some colours.

On-glaze decoration is applied and fired into the surface of the glaze after the glost firing; so, items which have been on-glazed will have been fired three times: biscuit, glost, and 'burn-on'; the burn-on not being at such a high temperature as the glost firing so that the colour range is much greater. Most expensive or ornate articles with an on-glaze decoration may have been fired many times to bring up the colours.

A gold or silver-edge line or decoration is normally on-glaze, a yellow line often being used for the cheaper under-glazing.

An on-glaze decoration can be felt on the surface of the glaze.

POINTS TO CONSIDER WHEN BUYING CHINA

When buying china, the following points must be considered.

1. On-glaze or under-glaze decoration?

This is important as few on-glaze decorations will stand the high temperatures used in washing-up machines; if the water temperature exceeds 60°C (140°F) and a detergent is used – as it will be – any coloured band or decoration will start to fade, but as long as no detergent is used, the temperature can rise to 93°C (200°F) before there is any risk of fading. Most washing-up machines wash at 60°C and sterilize at from 76 to 82°C (170 to 180°F) so this obviously limits the choice to an under-glaze decoration unless the manufacturer is prepared to guarantee, as some are, that the colour will not fade. To minimize the risk of colour fade, a neutral detergent is used.

2. The weight

In the trade there are only two types, *domestic weight* and *hotel weight*.

Domestic weight is the lighter with the body of the plate tapering to a thin edge (*see* Figure 3.1).

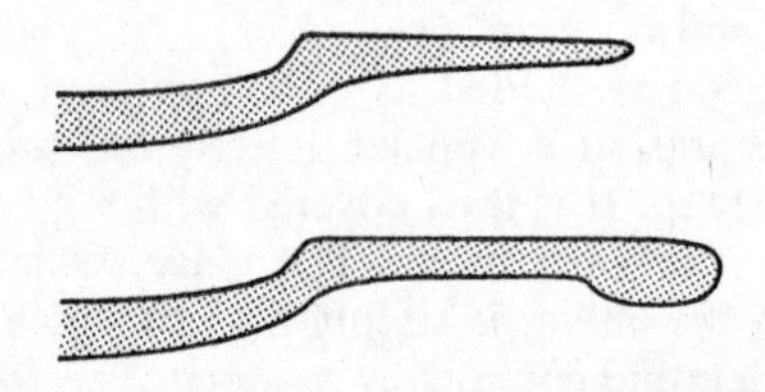

FIGURE 3.1 A tapered and a rolled-edge for china

To attract buyers in the domestic market, cups, teapots, and jugs are designed to draw customers and not primarily to give long life. Some manufacturers are producing a metallized bone china which is extremely tough and durable, and of domestic weight. Domestic weight china is often found in the smaller establishment where the washing-up is in the charge of one or two people and can easily be controlled. Hotel weight china is heavier with the body of the plate finished with a rolled edge. A rolled-edge plate can also be of domestic weight with a rolled edge added, but a hotel weight rolled-edge plate is ounces heavier in the body with the edging an integral part of the plate. This makes it much less vulnerable to chipping which, when it does occur, is under the edge of the plate and is not seen from above. Hotel weight stands up to heat from the hot plate or hot cupboard much better than domestic weight, without cracking or crazing of the glaze although a continuous subjection to changes in temperature does seem to lower the toughness.

Manufacturers test for the resistance to temperature change by heating the article to 177°C (350°F) then plunging it into cold water; this is repeated five times and is considered to be equal to two years of service. Resistance to chipping is tested by striking the edge with a swinging pendulum, the blow heavy enough to chip or break the plate being measured in foot-pounds. The more porous the body of the plate the easier it will chip or crack.

3. Shape and design

There is an increasing tendency to buy a style of crockery in which all items have dual use. In many cases, side plates can be used as under-plates for coupes, and butter and sugar or cereal bowls used with a side or under plate for soup. Where china teapots and hot-water jugs are used, lids should be interchangeable and easily replaced as a separate item. Jugs, hot-water jugs, and teapots should have a wide enough opening so that they can be cleaned easily inside. Short compact spouts are less likely to be chipped than the longer variety; all spouts must pour well and not drip. Handles should be welded firmly to the cup or pot and should be big enough to be picked up easily. Recesses in the saucer should fit the cup base.

The best design for a cup seems to be the straight-sided 'Worcester' type; these pack well in trays and, if the cups are rattled or knocked together, the point of impact is on the straight side and not on the rim. This also applies to an incurving edge (*see* Figure 3.2).

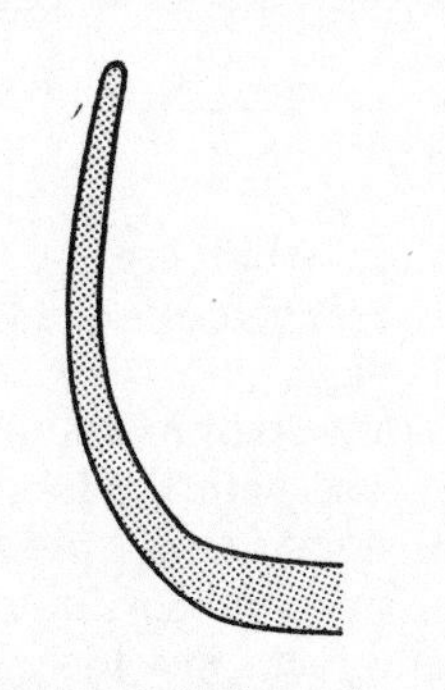

FIGURE 3.2 An incurving edge for cups and glasses

China has to be stored away easily and should be capable of stacking in piles of 25 or 30; stacking qualities should apply not only to plates and saucers but also to cups, bowls, teapots, and jugs. Unglazed foot-rims on plates or cups will eventually mark any glazed surface that they touch or rub against, but if the edge or rim of each item is the stacking point this is eliminated and will also make for a firmer stack.

4. The hardness of the glaze

A good hard glaze should not scratch; this can easily be tested and is particularly important when serrated-edge steak knives are in use; these can be ruinous to a soft glaze.

An alkali detergent can also attack a soft glaze, which is another reason why a neutral detergent is used for washing-up.

5. Colour and decoration

This should tone with the decorative scheme for the restaurant but should also be such that it will be suitable for any subsequent decoration as it is not always practicable to replace the complete range of china as frequently as the decorations.

Badges or one-way designs look effective but do make table laying and serving more exacting as all the china must be laid with the pattern the same way.

6. The size of plates and dishes

This greatly affects the portion control and can have a long-lasting effect on the kitchen budgeting; a good-sized portion of meat or fish on a 230 mm (9 in.) plate will look totally inadequate when served on a 280 mm (11 in.) plate.

7. Ease of replacement

The ease of replacement and the possibility of any future delivery delays.

8. Cost

The cost – whether the quality bought is of a suitable type and standard for the establishment.

BREAKAGES

These depend very much on the methods of washing-up and the care exercised by those handling the china; but arrangements can be made by the supervisor to minimize loss and damage.

When plates are correctly cleared in the restaurant or dining-room, they are taken to the wash-up already stacked with the food debris on the top plate. This means there are no haphazardly piled plates which can knock over and the clearing of swill and sorting are simplified. There should be sufficient space and racks in which to place the china prior to washing.

When washed, the china should be stacked and transported for storage on trolleys or trays and not carried by hand. Cups are either stacked or stored on trays so that, when they are to be used, a tray with 20 or 30 cups can be picked up from the storage cupboard and taken to the dining-room. There should be sufficient stock that staff do not have to go continually backwards and forwards with odd items in their hands. Trays or trolleys should be used.

A hotel report states that on average a cup is replaced three times in a year and a saucer once a year. The King Edward's Hospital Fund report on Crockery Washing, in a survey in 37 hospitals, gave the replacement figure for each person with decentralized washing-up as 3 plates, 2½ glasses, 4½ cups, and 2 saucers in a year. They found that with centralized washing-up the costs were reduced.

STOCKS

Stocks depend on peak periods but are usually 2½ times the amount required in use plus a 10% float in the restaurant.

GLASSWARE

The chief ingredient of glass is very pure sand to which is added other chemical ingredients to give glass its particular properties.

SODA-LIME GLASS

Soda-lime glass is made from sand with the addition of fine soda ash and limestone. This is used for the ordinary cheaper types of ware.

LEAD-CRYSTAL GLASS

Lead-crystal glass has red lead and potash added, to produce a slightly softer result with more brilliance and lustre than soda-lime glass. Because of its softer qualities, it can be cut easily and is used for cut-glass and crystal wares.

BOROSILICATE GLASS OR HEAT-RESISTING GLASS

Borosilicate glass or heat-resisting glass is a comparatively recent development having been manufactured in Germany towards the end of World War I and then further developed by the Americans who were the first to bring it into domestic use as ovenware and toughened glass. This process involves the addition of small amounts of borax to the sand which has the effect of reducing the expansion rate of the finished glass.

Normally, glass is a very poor conductor of heat. When heat is applied, the outer skin of the glass expands but the inner layer remains unchanged with the result that the glass cracks and breaks; thick glass will break quicker than a fine thin glass.

Ordinary glass is made by heating the ingredients to 1300°C or more and to 1600°C for heat-resisting glass. The molten glass is shaped either by direct mould pressure, pressed glassware which is the cheapest form, or by 'blowing' with the air pressure being obtained by hand or by machine. When moulded glassware is cooled, further heat treatment is often given to remove rough edges and to give a better, more polished finish. Depending on its composition, the glass will be transparent, coloured, or opaque.

Heat-resistant or toughened glass has a further treatment in which the glass, after shaping, is again heated to just below softening point, around 620°C, and then gradually reduced to room temperature. It may also be 'tempered' by re-heating to softening point and then cooling both outer surfaces rapidly, so making them shrink towards the centre of the glass. The ware is then re-heated to temperatures

FIGURE 3.3 Some different types of glasses

higher than it would meet in domestic use and passed through a cold 'waterfall'. This makes the glassware exceptionally strong with a chipping and breaking resistance far beyond that of normal glass. Most manufacturers claim that it is virtually unbreakable in normal use and that it will not crack, craze, or discolour. In some cases, the glassware will carry a guarantee for a year.

Although toughened glass does not break easily, some of it does disintegrate, and it has been known to 'powder' either in a storage cupboard or, more disconcertingly, in the washing-up water.

Heat-resistant glass had the disadvantage of being thicker and heavier than ordinary glass but it is now being made in 'thin-blown' varieties.

To avoid breakages the rim, which is the most vulnerable part of any glassware, should be slightly incurving so that the point of contact will be at the side and not at the edge. Sudden extremes of temperature should be avoided.

The shape and size of glasses varies according to custom and fashion and with the type of beverage to be served.

The most common shapes are shown in Figure 3.3.

The normal size for a sherry glass is 57 to 70 ml allowing 45 ml to the glass. The glass for white wine is usually 150 ml to give a 120 ml serving; red wine should be served in a larger glass, 170 ml. A water glass is usually of 180 to 225 ml capacity.

OTHER TYPES OF TABLEWARE

PLASTIC COMPOUNDS

These have been available since the 1950s; unfortunately they did not give the wear that was expected and were found to mark easily, scratch, and stain, and to have a 'plastic' taste. This reputation has taken a long time to disappear.

The modern plastics now on the market seem much more durable and stain-resistant. Most are made from melamine which is one of the thermo-setting plastics which becomes permanently hard under heat and the pressure of moulding. They now have a hard gloss surface which is more scratch and stain-proof.

Their great advantage is in their lightness, which makes them particularly suitable for use by young children and the very elderly, and in their non-breaking qualities.

WAXED PAPER OR 'DISPOSABLES'

These are widely used in vending machines but also have their place in any establishment when there is a sudden shortage of staff or when it is uneconomical to bring staff on duty to wash-up after late evening teas or coffee.

Disposable containers were originally made from waxed paper but are now also available in plastic-covered paper. Polystyrene is used for heat-resistant cups and bowls so that they can be handled easily without the need for a holder.

Sizes, shapes, and colours can be very varied.

The main problem devolving from the use of all disposables is what happens to them afterwards. Paper sacks or bins should be provided so that they are not left about the building.

WASHING-UP

About 90% of all damage and breakages to crockery and glassware occurs in the washing-up process and in the subsequent handling, stacking, and storage.

The main purposes of washing-up are:

1. To remove all debris and stains from plates, cups, and cutlery.
2. To sterilize and leave them free from all bacteria.
3. To produce clean, shiny, spot-free, and smear-free utensils and plates.
4. To do it in such a way that all crockery is intact and in a chip-free, crack-free, condition.

Washing-up is either done by *hand* or by *automatic or semi-automatic machine* but in all cases the procedure is the same, that of scraping off debris, sorting, washing, rinsing, sterilizing, and air-drying.

Hand-washing is still carried out in many of the smaller establishments; and is satisfactory provided that the two-sink method is used and the water temperatures are right (*see* Figure 3.4). Temperature gauges can be fitted to each sink.

After scraping and sorting, crockery is washed in the first sink at a temperature of 60°C (140°F). This is far too hot to immerse the hands, so rubber gloves and mops are used. 60°C is the recommended temperature but most washers-up prefer a lower temperature and use a dish-cloth; provided that the dishes are sterilized this is normally considered acceptable. When the water becomes dirty and greasy it should be changed completely; it is not advisable to 'top up' with more hot water. As a rough guide, 45 litres (10 gallons) of water and a detergent are sufficient to wash 100 to 120 plates. The detergent used should either be neutral or have an alkalinity value not above the value of pH 10·2; beyond this there is a risk both of injury to the skin for those who regularly handle the washing-up and of damage to the surface of the china.

After washing, the plates and utensils are packed in wire baskets for immersion in the sterilizing sink; this should be at a temperature of 76 to 82°C (170 to 180°F). Plates should be packed so that no two are touching and all surfaces are in direct contact with the sterilizing water. Plates and cutlery should remain in the sink for at least one minute (some authorities say two minutes), removed, and allowed to air-dry.

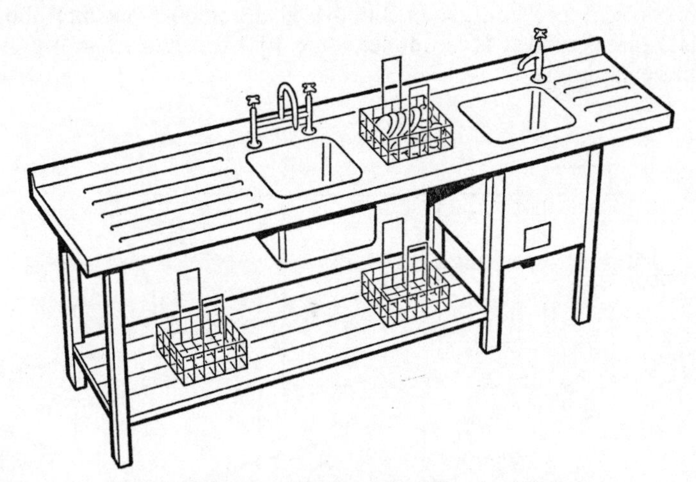

FIGURE 3.4 Combined washing-up and sterilizing sink unit

When the plates and utensils are completely dry, they should be stacked and stored for re-use. Tea-towels should be completely unnecessary.

Two people on hand washing-up should be able to cope with about 450 pieces each hour.

Where two sinks are not available, a *detergent/sterilant* should be used which will sterilize the dishes as soon as they are immersed. Soap, in contact with a sterilant, will destroy any bactericidal effect either possesses.

AUTOMATIC AND SEMI-AUTOMATIC WASHING-UP MACHINES

Automatic and semi-automatic washing-up machines (*see* Figure 3.5) are of three main types which are grouped in the way in which they bring the detergent solution in contact with the dishes.

They are:

1. By spray or jet – semi or fully automatic.
2. By turbulence – automatic.
3. By rotating brush – semi-automatic.

The same principles of washing and sterilizing apply. All operators should be fully instructed in the use of the machine, the temperature control, detergent quantity and control, and on how to clean the machine after each operating period.

All the machines are as efficient as the operators using them.

A small semi-automatic machine should be able to handle 2 000 to 3 000 pieces per hour, a medium-sized automatic machine about 6 000 pieces, and a large model some 10 000 pieces or nearly 300 racks each hour.

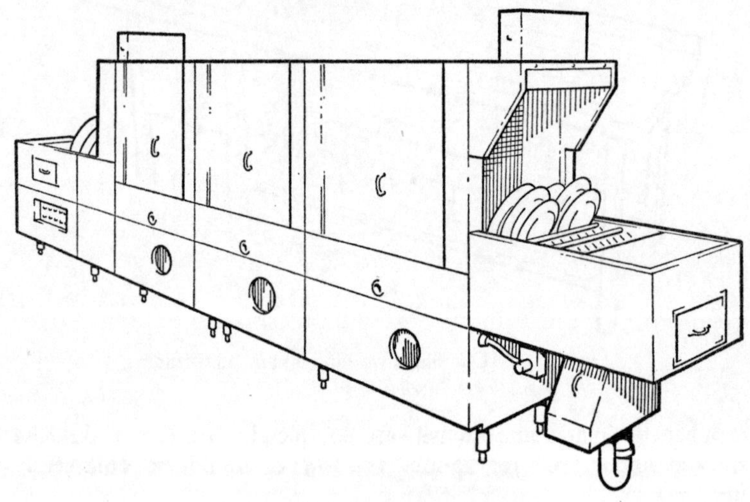

FIGURE 3.5 An automatic dishwasher
Width of conveyor 530 mm
Overall length 4·0 to 5·5 m

If the water used has a hardness of more than 7° a water softener must be incorporated in the supply to get the best results; this is particularly necessary with spray or jet-type machines as these quickly build up a chalk deposit which will block the jets.

Machines designed for glass washing are available and are suitable for ordinary glassware. Lead-crystal glass, as shown in tests made by the British Glass Industry Research Association, should not be repeatedly washed in automatic machines with detergents and very hot water as this can destroy the surface. Good crystal should be washed in warm water with a detergent, rinsed, and dried by hand.

The most common faults in washing-up are:

1. The failure to strip plates properly so that water and the washing compartments become clogged with debris. With some machines,

egg and mustard have to be rinsed off before the dish is machine-washed.

2. The operator is too quick in removing and stacking plates so that they do not completely dry. When stacked, the plates are found to be still wet and have to be towel-dried. This can also lead to brown stains on cups and plates which can be difficult to remove.
3. In hand-washing, plates have not been in the sterilizer for the required time; this means they will not be hot enough to air-dry completely and will not be sterilized.
4. Sterilizing is often forgotten in the evenings or at other times when there is little supervision or when part-time staff are on duty.

4 HOUSEHOLD LINEN

HOUSEHOLD linen is a very wide term which nowadays includes articles made from cotton, terylene, nylon, and wool, as well as those made from the traditional flax.

POINTS TO CONSIDER WHEN BUYING

SHEETS AND PILLOWCASES

For comfort, sheets should be big enough to tuck under the mattress firmly so that they do not come out during use. There should also be sufficient sheet to turn over and protect the blankets and quilt so that they are protected from dirt and soiling. Allowance must also be made for shrinkage; some cotton sheets will shrink by 5%.

Single sheet size of 203 × 274 cm (78 × 108 in.) and double sheet size of 228 × 274 cm (90 × 108 in.) are adequate and will fit most single and double beds apart from the 'king-sized' extra long beds which require sheets 274 × 297 cm (108 × 117 in.) in size.

Pillowcases should be at least 50 mm (2 in.) larger than the pillow in all measurements; the standard size is 508 × 762 mm (20 × 30 in.) and the size for a French or square pillow is 686 mm square.

The traditional material is *linen* which is an extremely hard-wearing and strong material, particularly when wet. Linen has the disadvantage of feeling cold to the touch and of crumpling easily so that, to retain a good appearance on the bed, the sheets have to be changed frequently.

Quality is judged by the evenness of the threads and by the closeness and smoothness of the finished weave. Linen sheets are expensive and, therefore, are not normally found in hotel or commercial use.

Cotton

Cotton sheets are not as expensive as linen and are not as hard-wearing, but they are much warmer and more comfortable in use and more crease-resistant. Some cottons may shrink by as much as 5%.

Union

Union is a mixture of cotton and linen and is used to give the best qualities of each: the linen to give a smooth finish and durability, and the cotton to take away the extreme coldness.

Flannelette cotton sheets

Flannelette cotton sheets are made from brushed cotton and are much warmer than cotton or union but without the same crisp appearance; because of their extra warmth they are often much appreciated by older people and by those from warmer countries who notice the colder climate.

The 'life' of a sheet depends very much on the care it gets in use and on the treatment it receives at the laundry, but it should survive 200 to 250 washes which, at the one clean sheet a week allowance of many college hostels, means a life of up to ten years. Most reputable manufacturers will guarantee their sheets for five, seven, or ten years, depending on the quality.

Of the newer fabrics, *nylon* sheets are very hardwearing but can be slippery in use and tend to be hot in summer and cold in winter, although brushed nylon has the same advantages as flannelette. Because nylon sheets slip and do not stay in place, they are often sold as fitted sheets.

Terylene/cotton sheets

Terylene/cotton sheets combine the hard-wearing, crease-resisting properties of terylene with the warmth and softness of cotton. Looking like a fine cotton, the sheet washes and dries easily and requires no ironing as any creases will fall out when it is smoothed into place on the bed. The sheets are finished with a soil-releasing agent so that the sheet does not easily pick up dirt.

Both nylon and terylene/cotton are more expensive than cotton or union but, if laundered on the premises, only require washing and drying and not the finishing which takes time, labour, and equipment. When they are sent to a commercial laundry, charges may be higher than the normal 'flat' rate as they need sorting apart from the cotton or union sheets and require a different washing process.

Most man-made fibre sheets carry a ten-year guarantee for wear but as they have appeared so recently, wearing comparisons between them and cotton and union sheets are difficult to make.

When buying

(*a*) If in quantity, obtain samples and test for shrinkage by sending to the laundry and comparing size and shape.

(*b*) With natural fibres, watch for the amount of dressing; with a cheap quality, extra body is given to the fabric. Rub the material and see if there is any starch 'fall out'.

(*c*) Look for a close smooth fabric.

(*d*) Hems should be firmly stitched. Sheets for domestic use frequently have a broad hem for the top of the sheet turn-over and a narrower hem at the base; this is not so usual for large-scale use as more work is involved for the room-maids and the sheet has more even wear if it can be used either way. Selvages should be firm.

(*e*) Pillowcases should be of 'housewife' tuck-in type and not have button or tape fastenings as these become detached in laundering and take longer to fit.

(*f*) Compare prices and samples and buy as good a quality as possible. Linen stored correctly will keep indefinitely so it is often possible to take advantage of better discount rates for quantity buying.

(*g*) Where there is a large order, manufacturers will mark linen by either embroidering the name of the firm on articles or by weaving the name into the fabric during manufacture. Where this is not done, linen is marked by machine-embroidery in the linen-room or labels are machined or pressed on. As a guide to the wearing qualities of each item, the date of issue is frequently added.

Linen Stock

This depends on the bed occupancy and on the service given by the laundry, whether the clean linen is returned within twenty-four hours

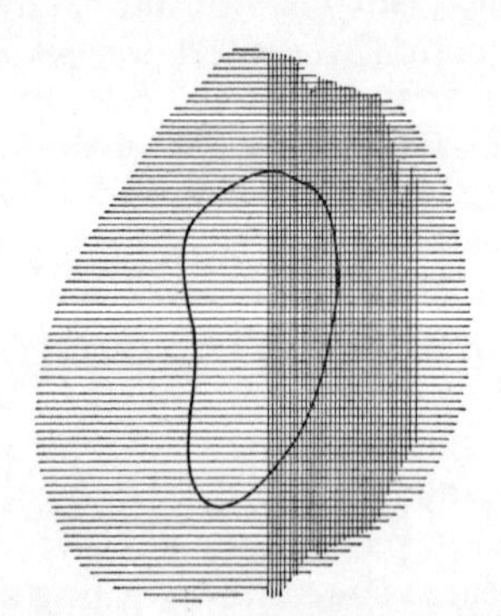

FIGURE 4.1 A machine-darn on linen
Work on the right side of material using a fine darning-cotton and needle.
The finished shape of the darn should be irregular to avoid strain on the material.

or in four or five day periods. A first-class hotel will re-sheet the beds daily whilst others may do so only every two or three days, whilst a college hostel may issue one clean sheet each week. Some colleges allow a minimum of three pairs of sheets to a bed with some managing on two pairs – although in this case there is no reserve when the laundry is delayed or where there are torn sheets.

Old sheets are used for dust covers or rags or remade for use as under pillowcases. Apart from mending small holes (*see* Figure 4.1) or 'hydro' tears in a strong sheet, there seems little point in wasting time 'edging to middling' sheets; when worn to that extent, the whole sheet will tear easily and be uncomfortable in use. (Hydro tears are caused through bad packing in the hydro-extractor at the laundry.)

Old pillowcases are utilized as underslips which protect the pillow, keeping the ticking clean. These underslips are laundered once or twice a year depending on use.

BLANKETS

In 1339, Mr Thomas Blanket was given permission by the Bristol Magistrates to make 'a fabric with a well-raised surface' for use as a bed-covering. The main development since then has been the introduction of open-weave cellular blankets which are warm and very light.

Blankets are made from:

Merino

Merino wool which is expensive but makes a soft, warm, and light blanket which wears well and is very durable if looked after properly.

Merino mixed with crossbred

Merino mixed with crossbred wool makes a heavier blanket, hard wearing and not so expensive.

Crossbred

Crossbred wool on its own gives a heavy blanket but one which is very serviceable and hard-wearing and is in general use.

Many wool blankets are permanently moth-proofed; even so, care must be taken in storage to prevent damage.

Most harm comes to wool blankets in the washing as bad washing causes the material to felt and shrink. Shrinkage is irreversible and results from friction between the ratchet-like scales on the surface of each wool fibre interlocking together. For this reason, low-speed washing-machine action is necessary, or else the blankets should be washed by hand. Provided the laundry is careful in the timing and speed of the rotary washers, no shrinkage should occur.

When washing, the machine runs for periods of one minute with a resting period of two minutes; for rinsing, the machine runs for half a minute periods with a rest period of half a minute.

Shrinkage will also occur if there are extreme changes of temperature in the washing water or if the material has been overstretched in the manufacturing process.

Many establishments prefer to wash their own blankets during summer vacation periods or else they are sent to be dry-cleaned; there is little difference in cost between this and the cost of the commercial laundry.

To offset the difficulties of laundering, blankets are now made from some of the man-made fibres.

Acrilan, nylon, and terylene

Acrilan, nylon, and terylene are used either on their own or with a percentage of wool. This makes for a blanket which is easy to wash, quick to dry, and shrink-proof. They are more expensive than wool but are more likely to remain in good condition for a longer time.

Hem-stitched edges should be firm; for a luxury finish a satin binding is used but this is more expensive and difficult to replace when worn.

Blankets should tuck in adequately but are shorter in length than sheets as they are not required to tuck in at the top. Single size of 175 × 250 cm (70 × 100 in.) and double size of 228 × 250 cm (90 × 100 in.) will fit all beds apart from those which are extra long.

Three blankets are usually needed but where there is central heating only two would be required; but extra blankets should be kept in reserve for particularly cold spells or where there are visitors from overseas.

Underblankets are needed both to protect the mattress and to make the bed warmer for the sleeper. These are bought to fit the mattress size or, more usually, old blankets are used.

BED-SPREADS

Bed-spreads can be made from a variety of fabrics but must be easily washable, crease-resisting as beds are invariably used for putting things on and for sitting, not too heavy in weight, and of a good colour and appearance which will tone with the general furnishing scheme.

The most usual fabrics are candlewick, folkweaves, terylene, man-made fibres, and most curtaining materials. A number of students' residences use coloured blankets which combine extra warmth with serviceability.

Bed-speads may be 'throw-over' or fitted but both types should reach to within 25 mm (1 in.) of the ground to cover the mattress base, and should have rounded corners so that the cover does not trail and collect dirt.

Throw-over covers may be extra long to allow them to be tucked in under the pillow to give a smooth neat finish.

Fitted covers are made for each individual bed and are not always interchangeable; the fabric used should either be shrink-proof or else washed before the cover is made. Sides can be either straight-pleated or frilled. The material frequently matches the curtains and other covers to give a uniform appearance to the room.

Eiderdowns or Quilts

Eiderdowns or Quilts are often provided for a warm light covering. Originally filled with down from the eider duck or with curled feathers, those with a terylene filling are now in general use and are more practical for large-scale use as they are easily laundered.

The underside of the quilt should be of a non-slippery material, such as cotton sateen or brushed nylon, so that it remains in place on the bed.

There is an increasing tendency to use larger down-filled quilts as a covering to replace blankets as is done on the continent.

TURKISH OR TERRY TOWELLING

Turkish or Terry Towelling was first introduced into this country in 1851 at the Great Exhibition in the Crystal Palace. It was called Turkish towelling having been discovered in one of the Sultan's palaces in Constantinople; it is also called terry towelling which is the old English name for any fabric with a pile of uncut loops.

The main requirements of the material are that:

(*a*) It should absorb water. The towel should be able to take up at least its own weight in water; most will absorb up to 125% of their original weight. The heavier and closer the weave, the more absorbency there will be.

(*b*) The towel should have a strong base weave with the loops firmly fixed so that they do not 'pull' easily.

(*c*) It should retain its colour.

When buying, one should:

1. Look at the finish to see that hems are securely stitched and that selvages are firm.
2. Feel the softness. A towel that is very soft is not likely to become rough with use but a harsh towel will become harsher.
3. Judge the weight for absorbency.
4. The heading or plain weave at each end should not be too wide; if it is, one is paying for decoration at the expense of the terry pile.
5. Test for shrinkage. A cheap towel may have been over-stretched.
6. Compare different samples for both size and price.

Sizes can vary considerably from one manufacturer to another; this accounts for the wide variation in price quotations often encountered.

Guest towels are approximately 380 × 610 mm (15 × 24 in.).

Hand towels vary from 559 × 965 or 1 016 mm (22 × 38 or 40 in.) to 610 × 1 118 or 1 168 mm (24 × 44 or 46 in.).

Bath towels from 762 × 1 372 mm (30 × 54 in.) to 1 016 × 1524 mm (40 × 60 in.).

Bath sheets from 1 016 × 1 778 mm (40 × 70 in.) up to 1 524 × 2 337 mm (60 × 80 in.).

Small hand or face towels are also made from linen, cotton, or union in a specially woven tough weave known as huckaback. These are placed in a bedroom for use in addition to a bath or hand towel of Turkish towelling, but their use is gradually dying out.

Bathmats

Bathmats vary greatly in size and fabric but are usually of a heavy cotton, Turkish towelling, or candlewick. They must be easily washable and be changed frequently. In many of the better hotels, freshly laundered bathmats will be placed for each guest.

When mats are of cork or a rubber material, they should be washed over daily by the room-maid.

A good size is 610 × 914 mm (24 × 36 in.).

Lavatory towels

Lavatory towels should also be changed daily. These are bought with 'toilet' woven across the centre, should have a taped loop for hanging up, and should be used for no other purpose.

They are normally made from terry or huckaback fabric but occasionally are cut down from larger articles by the linen-room.

Size is approx. 360 × 460 mm (14 × 18 in.).

Paper towels or tissues are frequently used in place of fabric towels.

Household towels

Household towels are best made from linen as this is a more hard-wearing fabric. Cotton or union is more water-absorbent but will leave fluff or lint on glasses and china. The material is often bought by the yard and made up in the linen-room.

Size is approx. 560 × 810 or 660 × 910 mm (22 × 32 or 26 × 36 in.).

Household towels for kitchen and washing-up use are rapidly being superseded by the increased use of paper towels and of steam-drying for crockery. This is advantageous on both hygienic and economic grounds as the cost is usually less than the cost of laundering.

TABLE LINEN

Table Linen is traditionally of a linen woven in a damask weave which is produced in a variety of patterns: ivy leaves, swans, roses, and geometrical designs which show by the reflection of the light. Serviettes or napkins should be the same pattern as the cloth.

Square cloths are 914, 1 372, 1 600, and 1 829 mm (36, 54, 63, and 72 in.) and, on a well-dressed table, should hang to a foot from the ground.

Oblong cloths are 1 320 or 2 290 × 1 830 mm (52 or 90 × 72 in.).

Napkins are 460 to 610 mm (18 to 24 in.) square.

Man-made fibres are being increasingly used for their crease-resisting, stain-resisting, and drip-dry properties, so reducing costs.

PILLOWS

The most usual size is an oblong of 482 × 736 mm (19 × 29 in.) and a square or French pillow 685 mm (27 in.) square.

The most usual filling is of *small feathers* with the hard centre spine removed or of *down* from the breast of the eider duck; these latter are very soft but expensive and would only be found in the most exclusive hotel. Both types are vulnerable to attack from moth or vermin if they are stored carelessly.

Kapok

Kapok, which is a fine cotton-wool obtained from certain seeds, is often used and is soft but in time tends to powder. Its advantage is that it is not attacked by moth; it is also used for asthma sufferers who are unable to sleep with a feather-filled pillow.

Terylene

Other fillings are made from *terylene* which is 'bulked' to give softness, and *rubber* or *plastic-foam.* In cheaper pillows, this plastic-foam may be granulated.

Feather pillows and those filled with kapok should be dry-cleaned or laundered every few years, terylene-filled pillows can be washed.

The same fillings can be used for *bolsters.* When these are used either separate bolster covers are needed or else the bottom sheet is tucked over.

Both pillows and bolsters are covered with a strong 'ticking' of striped or plain material which is protected by an under pillow-slip.

THE LINEN-ROOM

This is the control centre for all the linen and bedding used in the establishment and will also handle curtains and loose covers.

It should be situated where there is direct and easy access for loading and unloading baskets and for the collection and distribution of clean linen. A ground-floor room or one within easy reach of a lift is suitable. Linen baskets are heavy and it should be remembered that dropping these on the floor or working bench can have a disturbing effect on the ceiling and light fittings beneath.

Ideally, there should be two rooms, one for the collection and sorting of the dirty linen and the other for the storage and issue of clean linen. More often than not, this is not practicable and one room only is available; this should be sufficiently large to allow space for sorting, packing, and storing baskets or linen sacks. A diagram for the layout of a small linen-room is shown in Figure 9.5.

A good linen-room should:

(*a*) Be easily cleaned with walls and ceiling of a light-coloured washable paint or wallpaper.

(*b*) Be well lit.

(*c*) Have good ventilation and heating so that the laundry is kept well aired and mildew or dampness is prevented.

(*d*) Have strong, slatted, wooden shelves to allow the free circulation of warm air. Shelves must be well supported to take the weight of sheets and tablecloths. These are stacked in twenties or thirties. As a sheet weighs from 0·57 to 0·78 kilogrammes (1¼ to 1¾ lb.), the shelf load is several hundred kilogrammes. Shelves should be *labelled* to show each type of article.

(*e*) Have good security at both doors and windows with a counter or distribution table by the door to limit access.

(*f*) Be away from all kitchen and other smells.

The main equipment which is required is:

(*a*) Adequate working tables for inspection, sorting, and distribution of clean linen and for the cutting and making-up of curtains and covers, with a minimum size of 1 830 × 910 mm (6 × 3 ft). The table tops are usually covered with old blanketing and a sheet, but a coloured surface is better as it provides a contrast to the linen.

(*b*) A sewing-machine. In a large establishment, this will be either an electric or treadle machine with a 'free movement' attachment to enable the machine to be used for darning repairs and for marking. A smaller establishment might manage with a hand machine and do small darning repairs by hand.

(*c*) An iron and ironing table. A steam iron is of great use for removing creases from table-linen.

(*d*) Steps to reach higher shelves and storage cupboards.

(*e*) A telephone and account and record books.

The linen-keeper should be fit and active, as sorting and packing linen is a heavy and strenuous job requiring some physical effort. She should also be capable of controlling accurately the numbers of items in use and at the laundry.

LINEN EXCHANGE

This should be at a set time with the room maids exchanging on a 'clean for dirty' basis as a means of control to minimize loss and misappropriation. In most hotels, a small number of spare sheets, pillowcases, and table-cloths are kept in each department to allow for the emergencies which can occur when the linen-room is shut.

Room-maids should have a linen-basket, trolley, or canvas-lined floor basket to carry the linen. Dirty linen should be folded by the maid ready for checking and not bundled haphazardly.

Sorting and Listing

Dirty linen is sorted and packed to 'type' into hampers or canvas bags. These are usually provided by the laundry. Dirty linen should not be tied into bundles using a sheet as a container as these are invariably thrown about and will cause damage to the outer sheet.

Badly stained articles are sent separately or, if possible, the stain is removed before the article is sent.

Torn or damaged items of linen should be repaired before laundering to prevent further damage but, as no one likes to mend soiled linen, repairs are more often done when the laundry is re urned.

Linen must never be left damp in a warm atmosphere as mildew can develop quickly. This applies equally to laundry which has been returned from the laundry in a warm, not quite dry condition, and has been left over a week-end in the baskets before unpacking.

Laundry is listed in either duplicate or triplicate with one copy retained by the residence and the others sent with the laundry (*see* Figure 4.2).

When the clean laundry is returned, it is again checked and any differences are noted and reported to the laundry.

Before the linen is re-issued, it should be inspected for tears or worn parts; but some consider that, because of the time involved for the linen-room staff, room-maids should be able to check all linen issues as they are used and return any item for replacement if necessary.

All new linen *must be marked clearly* before it is sent to the laundry for the first time.

Mildew

When this is discovered early enough before the marks become fixed, the fabric can be washed with soap, left in a soapy condition, and dried in direct sunlight; this process is repeated until the marks have gone. If this is not successful, the linen can be soaked in a little bleach, or a paste of lemon juice and salt is smeared over the stains and the linen is again left in direct sunlight.

Name of establishment .... Consecutive No ....

Date ....

Type of article	Sent	Number checked at laundry	Checked on return
Bath mats	22	✓	✓
Bath towels	22	✓	✓
Pillow cases	168	✓	−2
Sheets	174	✓	✓

Total number of articles 386

Cost £ 9·65

FIGURE 4.2 A laundry list

STORAGE

The linen-keeper will keep on her shelves the surplus linen, blankets, and covers that are needed for normal use. These are kept either in cupboards or upper shelves where they are readily obtainable and are usually covered by dust sheets or polythene sheeting.

Reserve stock

This is kept in a separate store-room or storage cupboard which is kept securely locked, as are all stores. All linen must be kept dry and covered so articles are normally left in their original wrappings for protection. This also helps, of course, at stock-taking time. Some linens and sheets may yellow with storage but will whiten when first laundered. Sheets and pillowcases are always laundered before use and a small proportion of stock may be kept ready for issue.

Blankets, eiderdowns, pillows and bedspreads should be protected from moth attack by wrapping up well and by means of a pesticide aerosol, and should be inspected periodically to ensure that they are still in good condition.

STOCK-TAKING

Linen stock is taken at least once a year but where there is a large stock the policy may be for it to be checked monthly or three-monthly.

All linen is checked at the same time to avoid 'borrowing' by one section from another and errors on linen exchange.

A count is taken of all dirty linen, linen at the laundry, linen in use, and that kept as an emergency reserve on each floor, and linen in stock. To make checking easier, linen is stacked with the folds all the same way as shown in Figure 4.3.

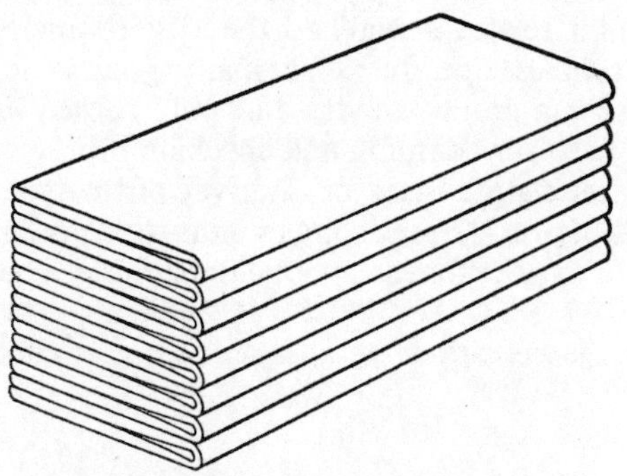

FIGURE 4.3 The regular stacking of sheets

Even so, all articles should be separated when counted as, if folded or stacked badly, one sheet or pillowcase can easily be mistaken for two.

A condemned book is kept by the linen-keeper as a means of writing-off all articles which are past redemption.

A stock sheet or book would be as shown in Figure 4.4.

Date ....

Article	Stock in hand (previous stock)	New stock	Total	Less condemned	Total	Actual stock	Discrepancy
Sheets	*918*	*36*	*954*	*29*	*925*	*919*	*−6*
Towels (bath)	*182*	–	*182*	*5*	*177*	*178*	*+1*
Towels (hand)	*598*	*60*	*658*	*34*	*624*	*613*	*−11*

FIGURE 4.4 Linen-room stock sheet

LAUNDERING COSTS

Establishments have the choice of using an outside commercial laundry, operating their own laundry either individually or in conjunction with other group establishments, or taking advantage of some of the linen-hire facilities which are offered.

Commercial or Group Laundries can charge or price in a variety of ways:

1. The most common seems to be a *flat-rate* price for each 100 items sent; each 'flat' article: towel, sheet, pillowcase, or serviette, is charged at the same rate. A separate agreed charge is made for overalls, blankets, bedspreads, and other 'odd' items.
2. A quotation based on the price per pound of dirty laundry. This is logical as it relates directly to the actual laundry cost as the washing-machines operate at so many pounds to each wash. Where there is a group laundry and well-marked linen, it means there is no need for counting and checking-in.
3. A monthly set charge based on a survey of the type and number of articles sent over the preceding months with any large variations costed at a fixed price for each pound or hamper – a bulk method of pricing.
4. A fixed charge per article with or without discount rates over a certain figure.

LINEN-HIRE

Linen-hire is steadily coming into greater use. Linen is supplied and laundered by the hiring firm at prices which compare favourably with the price previously paid to a commercial laundry for laundering charges. The linen-hire firm is able to do this by large-scale bulk buying, the standardization of linen, and the efficiency of modern laundry practices. The linen supplied is of a fixed and agreed standard.

An agreed basic stock is established and from this there is a straight exchange of linen two, three, or four times a week. The advantage is that no money is needed to maintain linen stocks or for replacements. Linen repairs are done by the hire firm.

A linen-keeper as such may not be required but there will still be the need for a responsible person to check the linen in and out and to distribute it to each room-maid.

The advantage or disadvantage of this method of supply can only be decided after a detailed analysis of the number and type of articles laundered over the preceding months + the average amount spent on linen costs and replacement costs + an estimation of the cost of repairs.

The convenience of linen-hire for roller-towels for cloakroom and kitchen use has been evident for many years.

OWN LAUNDRY

Whether it is economical or not to operate one's own laundry depends on the size of the establishment. One estimate considers that where 20 000 pieces of laundry are needed each week there may be sufficient cost incentive to make it viable. Careful costing for capital equipment, overheads, and staff are needed, particularly now that so much automatic or semi-automatic equipment is available. Now that non-iron terylene and cotton bed-linen and drip-dry fabrics are available, this means that part of the expensive laundry equipment – the calender or roller-ironer – is not required.

The basic equipment which is required comprises washing trolleys with built-in weighing, rotary washing-machines (Figure 4.5), hydro-

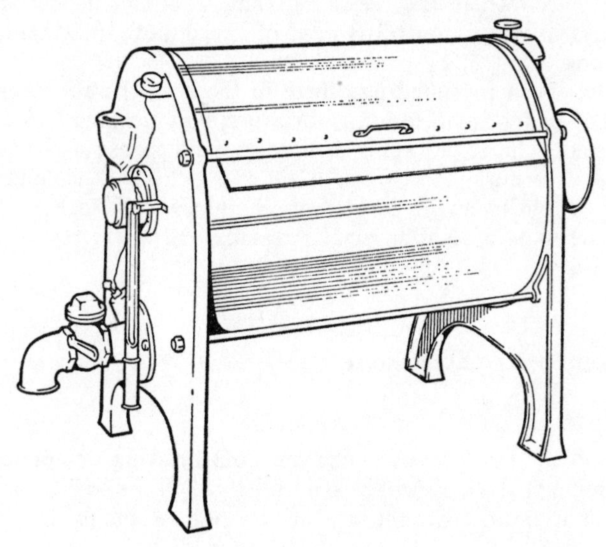

FIGURE 4.5 A rotary washing-machine of 22 or 27 kg (48 or 60 lb) dry weight

extractors, drying cabinets or tumbler dryers, and a small press or steam-iron for finishing.

The great advantage of home or group laundering is the speed of return of linen back to the linen-room which permits lower stocks and less capital involvement. Better control over the washing process is considered to give linen a 30% longer life.

Whilst a large number of firms would not consider doing all their own laundry, a considerable number possess a small washing-machine, spin-dryer, and drying cabinet, and wash their own blankets, curtains, and small items. Some also take advantage of local launderette facilities, particularly for towels and other items which require no finishing.

5 PURCHASING, CONTROL, AND COSTING

WHERE possible costs and comparative costings have been introduced at the places where they seem relevant. This chapter summarizes the ways in which control is kept of expenditure and methods of budgeting.

As has been indicated elsewhere in the text, there is a growing tendency to rent or hire linen, furniture, and cleaning and kitchen equipment. The advantage is that the capital expense is avoided and the cost of hiring and usage can be added directly to the running costs of the establishment. In most cases, maintenance is included in the hiring terms as is also the regular replacement of the equipment or furniture.

PURCHASING

There are two usual methods:

RETAIL

1. Retail for small items which are non-recurring or are required infrequently. These will be paid for through the Imprest or Petty Cash account. Discount may or may not be obtainable.

WHOLESALE

2. Wholesale for larger amounts and for frequent purchases. Goods are bought from wholesalers, distributive agents, or direct from the manufacturers: wholesale prices are lower than retail and, depending on the payment dates and quantity bought, will frequently carry a substantial discount. It should be noted that goods which carry purchase tax are often offered as the cost of the article plus the amount of purchase tax payable so that the two amounts have to be added to obtain the cost to the customer. This is done to simplify the calculation of discounts as, naturally, no discount is allowed on purchase tax. It also avoids the necessity to re-print catalogues when purchase tax rates are changed.

   Discounts and prices are negotiable either by the individual establishment or by a purchasing officer or committee on behalf of a number of different establishments or units in a group. This is bulk or contract buying.

CONTRACT BUYING

Contracts can run for any length of time. Some may be for a period of one to three years, whilst a contract for the supply of perishable goods may be for a shorter period of three to six months. It depends very much on the supply and demand of the article and on the price fluctuations which may be expected.

TENDERS

Contracts are put out to tender usually by placing an advertisement in the local or a national paper or by inviting firms to submit a quotation for the goods required: for example 'tenders are invited for the supply of soaps and detergents to the XXX Hotel for a period of two years from 1st April, this year. Further information can be obtained from the Accountant at the Hotel'. It is frequently stated that canvassing disqualifies.

Suppliers will need to know the approximate delivery dates and times for delivery, whether goods are dispatched to a central store or to the individual units, and the amounts required in each period.

Most firms insist on at least three tenders so as to obtain a comparison of prices.

Contracts should show: the commodity; the unit of size or weight; the quality standard; terms of payment, net, monthly account, discount, carriage paid, etc.; delivery, number of days from receipt of order, or ex stock; the address for orders and correspondence.

Contract prices are confidential and one should beware of the salesman who casually asks 'and how much are you paying for that?'

TOTAL USAGE

It is necessary to know the total amount used in each period when negotiating a contract. The simplest method is to use the completion of an average monthly consumption or usage sheet which shows the main items used for which contracts are made. This, totalled for the year with the totals from all other units, gives the average quantities for purchases for the group.

MONTHLY CONSUMPTION SHEET

Commodity	Unit of weight or size	Jan.	Feb.	Mar.	etc.	Total
Detergent	5-gallon drums	2	1	3	..	21
Polish – water based	1-gallon	1	–	–	..	6

Figures are taken from the stock books and are quickly added each month either by the stock-keeper or in the office. Besides being the basis for all contracts, these figures are useful as a quick check on increases (or decreases) in use of any item so that a decision can be made as to its justification. Increases should, of course, be noticed from the stock book and issue sheets but can, in practice, easily be overlooked. A further use of monthly consumption sheets is for calculating the effects of price increases; if, in the example given above, the price of detergent is increased by 10p a gallon this will mean an extra expenditure of £10·50 for the year if the same standard of detergent is to be used and the same rate of issue is maintained. This sort of information is essential when asking for or justifying an increase in budget expenditure.

STANDING ORDERS

Standing orders are placed for regular deliveries of goods; these state the quantity and dates when the commodities are required. This can lead to either overstocking or shortage and is only satisfactory when demand is constant. It may also mean that advantage is not taken of price reductions or 'good buys'.

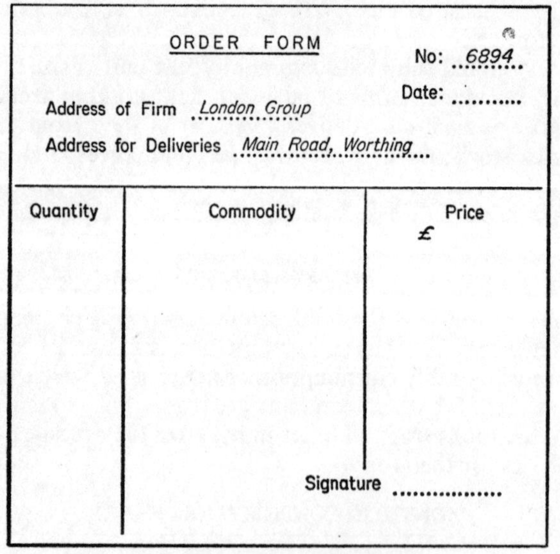
ORDER FORM

No: 6894

Date: ....

Address of Firm London Group

Address for Deliveries Main Road, Worthing

Quantity	Commodity	Price £

Signature ....

FIGURE 5.1 An order form

ORDER FORMS

Order forms (*see* Figure 5.1) are usually printed in duplicate but three or four copies may be required: one copy for the supplier, one copy to be attached to the invoice or statement of the goods received and passed for payment, one copy to the stores for information and for checking, and the fourth copy may be needed for office records. Orders can be given verbally, and often are by telephone, but written confirmation should be sent. The order form is numbered and dated, and states the unit and amount of the goods required, the price, and where they should be delivered, and it should always be signed by an authorized person.

ADVICE NOTES

Advice notes are normally used only if goods are dispatched from a distance by British Rail or by carriers. If goods are not received, the suppliers are notified within the time stated on the advice note.

DELIVERY NOTES

Delivery notes arrive with the goods and are checked and signed by the Store-keeper stating that the goods received are in good condition and correct.

STOCK BOOKS

On delivery, all goods are entered in the stock book. This should state:

Commodity	Detergent	*Price*	27 p per gal			*Unit*	5-gal
Date	*Received*	*Issued*	*Total*	*Date*	*Received*	*Issued*	*Total*
5.2.71	6	–	6				
18.2.71	–	1	5				
3.3.71	–	1	4				
					Minimum stock 2 × 5-gal		

Stock sheets show the price and price alterations when they occur with the date when the change was made. The minimum stock level should also be shown for re-ordering by the stores. Care should be taken to ensure that stock is used in rotation.

In a large establishment, issues are taken-off stock daily but, in a smaller unit, this may only be necessary each week. Stock should only be issued to any department on receipt of a signed indent which should be in duplicate: one copy is retained by the store, the other returned to the department with the issues. This is particularly important in large establishments as any stock discrepancies can easily be 'adjusted' by altering the issue sheets when taking-off stock; this is not as likely to happen when the other copy is held by another department. When necessary, departmental and stores issue sheets can be compared.

For the same reason, stock books should not be kept by the person who issues the stores; but this is, of course, ideal and will only happen in a very large establishment.

STOCK CHECKS

The best method is to take a physical check of stock and then compare the result with the totals as shown in the stock book. Again this is best done by someone other than the person issuing stock. Because it is easy to make inadvertent mistakes in booking receipts and issues, it is wise to spot check some items weekly. Although it is the policy of some firms to produce a stock figure monthly when the value of stock held is costed out, other firms may only require this to be done at six-monthly or yearly intervals. The figures needed to determine expenditure are: the amount spent + the value of stock at the end of the last accounting period − the value of present stock. This gives the expenditure for the period.

It may be the policy of the firm that the value of stocks held be limited to a percentage of the budget estimate; this might be fixed at 5 or 8%. It can sometimes be part of an Auditor's responsibilities to take stock periodically when the books are audited.

BIN CARDS

Bin Cards are very similar to stock sheets and are kept in all large stores. They are attached to the bins or stocks of goods and are a record of all receipts and issues. They are necessary when the storeman does not keep the stock book and can be very useful as a double check.

GOODS RETURNED BOOK

Goods returned book records the return of all goods sent back to the suppliers. Items which are returned are taken off stock in the usual way.

THE EMPTIES BOOK

The empties book records the receipt and return of all chargeable containers, tins, and sacks. Without some checking system, these can quickly disappear.

After being checked as correct and the items entered in the stock books, *invoices* or statements are attached to the *order form*, signed as being correct, and passed for payment.

EXPENDITURE FIGURES

In a small establishment, it may not be considered necessary to keep detailed figures of expenditure. In a large establishment it is indispensable as a means of controlling running costs within the company, as a means of comparison of costs between similar establishments, and as a basis for forecasting future expenditure.

Expenditure figures may be produced from within the department or may be prepared at set periods, six-monthly or yearly, by the finance department.

To help the preparation of these figures, all expenditure is coded to the right budget heading; that is, all laundry bills would be coded 01, Cleaning 03, Heat and Light 04, etc. Bills are normally coded before being passed for payment. In a group, each unit will have a code number which is placed in front of the budget heading code number, eg. 26/01. The figures received might look something like Figure 5.2.

As can be seen, some years will balance out rather better than others. Some of the points which might be raised by the finance department or departmental head are:

1. Should a supplementary estimate have been asked for in 1969/1970?
2. Do costs compare favourably with similar establishments?
3. Has the increase in provision costs been caused by a better standard of catering, higher food costs, increased wastage, or a greater number of residents than had been estimated for in 1968?
4. Has the standard of catering fallen in 1970/1971, have prices been more competitive, or has there been more control of receipts, issues, and waste in the kitchen?
5. Have staff become more productive in 1970/1971?
6. Should laundry costs have decreased from the 1969/1970 period?

Without figures of this type it is not possible to judge how efficient and viable is an establishment. In most firms opportunity is given for the revision of estimates during the current year.

Establishment .... Accounting period *April 1970 to March 1971*

		1969/1970					1970/1971				
		No. of resident days = 25 000					No. of resident days = 27 000				
	Code No	Estimate	Amount and % of total		Cost per resident day	+ or −	Estimate	Amount and % of total		Cost per resident day	+ or −
		£	£	%	£	£	£	£	%	£	£
Cleaning materials	01	400	382	1	0·01	−18	400	415	1	0·01	+15
Heat, light and power	02	6 000	5 850	13½	0·23½	−150	5 900	5 980	14	0·22	+80
Laundry	03	2 000	1 950	4½	0·08	−50	2 000	2 030	4½	0·07½	+30
Provisions	04	10 000	10 600	24½	0·42	+600	10 750	10 710	24½	0·40	−40
Wages etc.	05	24 000	24 400	56½	0·98	+400	24 500	24 545	56	0·91	+45
Total		42 400	43 182	100	1·72½	+782	43 550	43 680	100	1·61½	+130

FIGURE 5.2 Analysis and comparison of expenditure

ESTIMATES

Estimates are produced for both capital and revenue expenditure. Revenue expenditure covers the cost of the day-to-day running of the establishment and, besides the items already mentioned, will include maintenance, postage, telephones, printing and stationery, etc. Capital expenditure is normally regarded as being either items bought which have a 'life' extending beyond the financial year or an item costing more than £xx in value.

For ease in calculating the number of resident-days when both residents and non-residents are catered for, it is usual to consider the non-resident figures as a percentage of a full day (*see* Figure 5.3). The percentage agreed for each part of the day would be related to its value.

Where no charge or only a nominal charge is made for food and accommodation, this type of calculation and recording is necessary for accurate costings.

Non-Resident Meals supplied
Period ended ....

Meal	*Total Number*	*Points* (as % of a day)	*Total*
Breakfast	80	20	1 600
Coffee	1 000	3	3 000
Biscuit	1 000	2	2 000
Lunch	950	35	33 250
Tea	1 000	2	2 000
Cake/biscuit	1 000	3	3 000
Dinner	120	35	4 200
		100	49 050

Non-resident meals in terms of full residence would be:

$$\frac{49\,050}{100} = 490\tfrac{1}{2} \text{ residents.}$$

FIGURE 5.3 Record of non-resident meals as a percentage of full residence

INVENTORIES

In the same way as consumable goods are recorded in a stock book, items which are in general use and represent capital expenditure are entered in an inventory room by room (*see* Figure 5.4). Any alterations or replacements are entered as they occur. This is why most housekeepers prefer furniture to remain in its room of origin and do not like chairs and tables being moved around the building. A permanent move should mean an alteration in the inventory, as would furniture which is broken or replaced.

Inventory
Block *E*
Floor *1st* Room *45*

	Date checked								
Description		2/70	8/70						
Divan	1	✓	1						
Mattress	1	✓	1						
Dressing-table and mirror	1	✓	1						
Chair – easy	1	✓	1						
Chair – upright	2	✓	2						
Chest of drawers	1	✓	1						
Bedside table	1	✓	1						
Potable lamp	2	1 1(new)	2						
Rug 1·3 m x 1m (green)	2	✓	2						
Curtains – 1 pair	1	✓	1						
Initials	[illegible]	[illegible]	[illegible]						

FIGURE 5.4 An inventory

6 CLEANING

FACTORS AFFECTING COSTS

THE cleaning requirements for buildings can vary considerably with many factors affecting the ultimate costs and cleaning frequencies. Some of the main problems to be considered are:

1. The condition of the building

The condition of the building and the wall, ceiling, and floor surfaces. Some establishments have innumerable corners, crevices, and ledges which harbour dust; others have cracked and irregular plastering and uneven floors; some have large landscape windows which are cleaned more quickly and efficiently than windows with small panes of glass. All such variations affect the individual costs and time allowances.

2. The amount of dust and dirt which is brought into the building

The amount of dust and dirt which is brought into the building of which some 90% is carried by footwear or by wheeled traffic. This can be reduced by carefully designed entrances with adequate dust removal mats or sunken grilles and by ensuring that all traffic lanes into the building are paved and are kept clean by regular brushing, washing, or hosing down.

3. The degree of air pollution

The degree of air pollution will have a considerable effect on cleaning frequencies. It will be found that net curtaining used in an industrial area may need washing every two weeks whereas, if used in a country district, the washing frequency may be extended to two or three months or more.

4. The type of furniture

Fitted furniture is generally more labour-saving than that which is free-standing as dust cannot accumulate behind or under it and there is no need for it to be moved regularly for cleaning. Furniture which is heavily carved or with many ledges will require a longer cleaning time allowance than stream-lined furniture; the surface finish will also affect the time.

5. The type of floor covering

In older buildings this may be very varied and may involve several different cleaning processes in the same area, carpet squares with highly polished surrounds being notorious time-wasters.

6. The use to which the building is put

7. The availability, skill and training of the staff

8. The equipment available

9. The money available

As with so many other things, cleaning is a tug-of-war or compromise between the standards which it is desired to maintain and the amount of money which is available for an 'unproductive' service.

STANDARDS AND TERMS

Because conditions vary considerably from one establishment to another, it is difficult to compare costings accurately. On page 79 is shown how this can be attempted.

As there is, occasionally, a little uncertainty as to what is meant by a well-dusted table or a well-swept floor, it is as well to define the *standards* that one expects and some of the *terms* commonly used. These are:

Low dusting

A properly dusted surface is one which is free from all dust, stickiness, marks, or streaks. Dusting is to a height which is within the arm-length of the cleaner. Equipment includes damp dusters, impregnated gloves or mops, vacuum dusters, and the ordinary dry duster.

High dusting

This is the removal of dust from walls, ceilings, pipes, ledges, equipment, and fittings, which are normally beyond reach from floor level. Equipment includes light, flexible, or extending impregnated mops or vacuum cleaner extensions. It should be remembered that these can become tiring if used for any length of time.

Damp wiping

Should leave a surface free from all dirt, grease, and marks. The finish should be clear with no film or 'bloom' on the surface. Equipment used: a well wrung-out net cloth or chamois leather using either water, water and vinegar, or water and a little detergent. The surface is then dried or buffed up with a soft dry cloth.

Sweeping

A floor is well-swept when it is free from all dust and debris. Embedded or ingrained dirt or grease will not be removed by sweeping.

Equipment used: (*a*) impregnated sweeping mops either straight or v-shaped which extend to a diameter of up to 2 m (6 ft); this means that most corridors can be swept in one action; (*b*) vacuum sweeping; and (*c*) brushes used with a dust-arresting compound. Ordinary nylon or fibre brushes are of questionable value as they distribute dust, particularly into the air, which re-settles on a surface, and as they invariably leave fluff and bits behind them great care is needed in their use. The size of the equipment should be suitable for the area to be covered. Many sweeping 'heads' can be bought in sizes up to 760 mm (30 in).

Dry mopping

This is the light polishing of a floor using either a dry polishing mop or one which is lightly impregnated.

Damp or wet mopping

Should leave a floor free from all dirt, marks, streaks, and pools of water. Equipment used: a wet mop and bucket and detergent and water. If soiling is heavy, the detergent/water must be changed regularly, otherwise the dirt will only be re-distributed on the floor. Because water damages so many types of floor, the amount used should be kept to the minimum and it should be removed and the floor dried as soon as is possible. Notes on damp mopping a floor are shown in Figure 6.1.

Notes for Instructing Staff

HOW TO WET-MOP A FLOOR

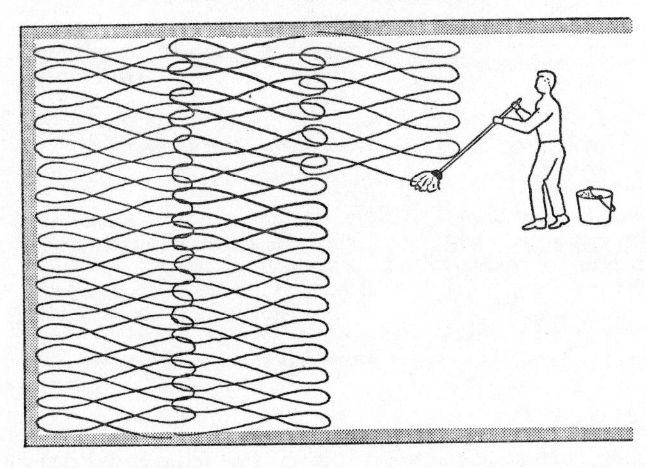

Used for the removal of slight soiling and dirt: it is *always* preceded by sweeping to remove dust and debris. [*contd.*

Equipment required: Sweeping tool, mop bucket with wringer attachment, wet mop, 16 to 32 oz size (suitable for room area), detergent/water solution.

What to do	*Key points to mention*
Assemble equipment	Care with detergent/water solution to prevent spillage; if required, put mat under bucket to protect other floor areas. Tidy equipment.
Clear furniture and floor coverings from area	Place tidily
Sweep floor	See notes for sweeping
Divide floor area into convenient work sections	A large room into 4 sections; a hall into ½ or ¼ width sections; etc.
Start mopping	Start at section farthest from finishing point to avoid walking over cleaned or wet part of floor. Put mop into detergent solution and wring hard. Run mop round room edges and into corners using the heel of the mop. Clean rest of section using a figure of 8 movement, each time just overlapping the previous movement. Do not splash skirtings. Rinse mop frequently; always wring hard to prevent excess water on floor
Inspect floor	Prevent others using floor until it is dry. Drying can be hastened by opening a door and windows to allow a cross-current of air through
Replace furniture and floor coverings	When floor is completely dry
Remove cleaning equipment, empty and clean, and place in cleaners' store	Empty and rinse bucket at cleaners' service point *only*, not at utility room sinks; dry and leave to air. Wash mop in clean detergent solution, rinse, wring out hard, and leave to dry

FIGURE 6.1

Scrubbing

A floor is well-scrubbed when it is left free from embedded dirt or grease, debris, stains, marks, and the remains of the cleaning detergent and water. Scrubbing must be followed by a clear water rinse and the removal or 'pick-up' of water. Again, the amount of water used should be kept to a minimum. Equipment used: detergent and water and a scrubbing machine or, for manual use, a deck-scrubber. Pick-up is either by the use of a wet vacuum, squeegee, or wet mop. Where dirt is very ingrained, it may be necessary to use a nylon or medium-grade metal fibre pad under the head of the scrubbing machine.

Dry or spray cleaning

With the use of nylon or metal-fibre pads attached to the polishing machine ingrained dirt, traffic marks, and streaks are removed. Whilst the machine is polishing the operator sprays either an emulsion or spirit-based wax polish on the floor; this mixes with the dirt and is absorbed by the pads which are changed when dirty, washed, and re-used. This process cleans and polishes in one operation; the skill lies in the correct application of polish which should be sparingly used.

Wax removal

Old polish and dirt which has built up in the polish is removed by the application of a stripping compound or 'stripper'. This loosens the polish which is absorbed by nylon or metal-fibre pads fitted to the polishing/scrubbing machine. This is followed by wet-mopping to remove any dirt or polish which has not been picked up by the pads. The floor is then scrubbed with detergent, water, and a neutralizing agent, to remove any remains of the stripping compound. This stripping compound is basically a strong alkali which, if not neutralized by an acid neutralizer, will continue acting on any subsequent polish which is applied to the floor. This may make the floor sticky and impossible to polish or give a patchy effect; the only satisfactory remedy is to strip off and start again. If a proprietary neutralizer is not available, a solution of vinegar and water makes a good substitute. The frequency of this treatment depends on the amount of dirt build up and the condition of the 'traffic-lane' areas of the floor.

Primary sealing and waxing

In many cases wax removal is followed by the application of one or more coats of a plastic seal (*see* page 86). The seal is applied with a clean applicator and allowed to harden before the area is used. Some seals dry bright and require no further polish but others need protecting with two or three coats of polish. These are lightly buffed between applications.

If no seal is applied, wax removal is followed by the application of two or three coats of either a spirit or water-based polish depending on the type of floor, the floor being lightly buffed after each application. For notes on polishes *see* page 88.

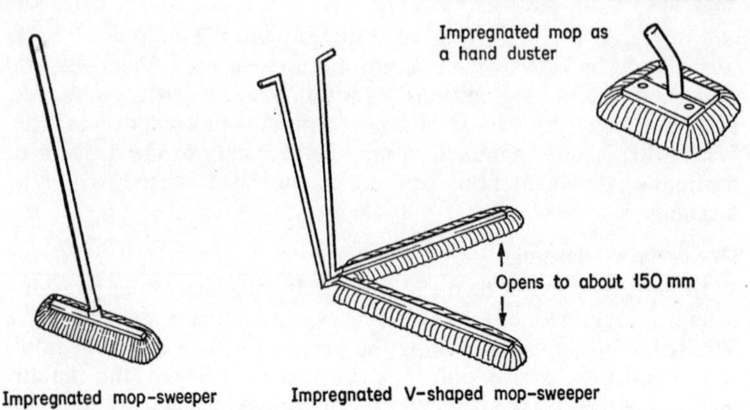

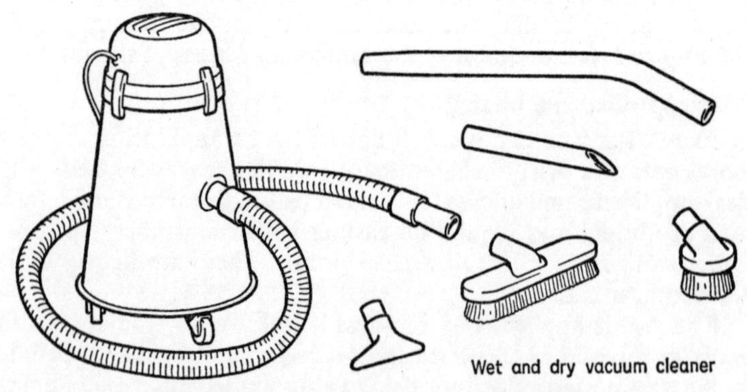

FIGURE 6.2 Cleaning tools and equipment

Touching-up wax

This is applied as it is required to heavy traffic lanes; these may require to be damp mopped before the polish is applied. When buffed, the floor should have an even gloss.

Polished metal

When clean, metal should have a bright appearance with no sign of tarnish or cleaning agent remaining either on it or the surrounding surfaces.

Light fixtures

Are clean when all parts, including the bulbs or tubes, are free from dirt, insects, and dust. As dirt frequently becomes baked on, some of the fittings may need soaking and washing. Care must be taken to ensure that the electricity is switched off before cleaning starts.

To add to the difficulties of maintaining standards, the findings of one Work Study team were that there is no known method of objectively measuring the polished condition of a floor or its cleanliness.

Common cleaning tools and equipment are shown in Figure 6.2.

CARE OF FLOORS: methods of cleaning

Type of floor	*Initial treatment when new*	*To clean and maintain*
Wood, hard or soft; strip, wood block, parquet or composition	Normally sanded, sealed, and waxed by contractor	1. Sweep with impregnated dry mop, or 2. Spray-clean and wax using a solvent-based polish, or 3. Wet-mop, using a neutral cleaner, use as little water as possible and dry at once, or 4. Wax and polish.
Cork	Normally sealed and waxed by contractor	As for wood *Avoid* the use of water
Linoleum	May be bought sealed or un-sealed; if un-sealed, the top dressing is removed before a seal can be applied	As for wood *Avoid abrasive cleaners* and alkalis and excessive water
Thermoplastic, P.V.C., vinyl, and asbestos P.V.C.	Avoid using water for 2–3 weeks after laying	As for wood *Do not* use a solvent-based polish

Type of floor	*Initial treatment when new*	*To clean and maintain*
Rubber	Avoid using water for 2–3 weeks after laying to allow adhesive to dry and harden	1. Sweep with impregnated, oil-free mop, or 2. Wet-mop, using a neutral detergent, or 3. Apply wax polish as recommended *Avoid* solvent-based polishes, strong alkali and hot water
Asphalt	Avoid using water for 1–2 weeks after laying	1. Sweep with dry mop, or 2. Wet-mop, using a neutral detergent, or 3. Apply wax polish as recommended *Avoid* solvent-based polishes
Terrazzo and granolithic	Normally sealed by contractors	1. Sweep with impregnated mop, or 2. Wet-mop, using a neutral detergent, or 3. If very dirty, scrub using an abrasive, and rinse well, or 4. Can be polished, but a high gloss is often obtained without using a polish by buffing with a floor machine *Avoid* acids and alkalis
Marble		1. Sweep with impregnated mop, or 2. Wet-mop, using a neutral detergent, or 3. If required, apply polish *Avoid* acids and abrasives
Ceramic and quarry tiles	Avoid using water for 1–2 weeks after laying	1. Sweep with untreated or impregnated brush or mop, or 2. Wet-mop, using a neutral detergent, or 3. Scrub
Concrete	Normally sealed by contractor; may be painted.	1. Sweep with floor brush or impregnated mop, or 2. Wet-mop

TO CLEAN CARPETS AND RUGS

Surface dirt should be removed daily using either a carpet sweeper or a vacuum cleaner. Deep, embedded dirt can only be removed with a vacuum cleaner. Carpets which become badly soiled should be shampooed either by hand or by machine.

The treatment of floor covering and carpets is discussed fully in *Services and Maintenance*, the companion volume to this book.

FREQUENCY OF CLEANING

Some cleaning has to be done daily, some weekly, and some will only be required at less frequent intervals. These may be monthly, quarterly, or even yearly, depending on the standards worked to and the degree of soiling.

Whatever the schedule, however, all staff must be trained to remove stains and marks as and when they occur so that they do not become ingrained and permanent.

WORK THAT IS NORMALLY CARRIED OUT DAILY

Floors, corridors, and skirtings	– All well-swept or dry mopped. Brush or mop skirtings with the floor
Furniture, all horizontal surfaces and fittings, telephones, and radiators	– Dust, stickiness and finger-marks removed
Waste-paper baskets	– Remove, empty into paper sacks or bins, clean, and return
Ashtrays; may require more frequent cleaning	– Empty, clean, and return
Metalwork	– If not lacquered, then clean and polish
Doormats	– Sunken grille type: vacuumed. Fibre: taken up and cleaned
Toilets, wash-handbasins, and bathrooms; may require more frequent cleaning	– Remove all waste towels and litter. Disinfect and clean toilets and urinals. Clean wash-basins and baths by removing debris from waste grilles. Wipe with a cloth and a mild abrasive, particularly all edges and crevices, rinse, and dry. Rub up chromium taps and fittings. Check and clean incinerator. Replenish soaps, towels, etc. Check floor; damp mop if necessary

WORK CARRIED OUT WEEKLY

Floors and corridors	– Polished floors buffed when required, touch-up wax applied and buffed when necessary
Furniture, wood, polished	– Clean with a cloth dampened with either water, vinegar/water, or detergent/water. Dry and polish
Upholstery	– Clean to remove dust, using vacuum cleaner and appropriate attachment
Paintwork, doors and window frames	– Wipe over with a dampened cloth. If in a heavily air polluted area, clean with detergent/water and rinse with clear water to remove all finger marks, smoke film, and grease

Walls – Dust, using clean wall brush or vacuum attachment, to within arm reach

Windows – If not cleaned under contract, clean using vinegar/water and chamois leather, scrim, or squeegee. Cloths are well wrung out; it should *not* be necessary to polish with a dry cloth

Mirrors and pictures – Should be kept as dry as possible to prevent damp seeping into the silvering or picture. Methylated spirits cleans and evaporates quickly and will also remove hair lacquer from mirrors, but is highly inflammable for general use; or use a dampened chamois as for window cleaning or a proprietary brand of cleaner. Oil paintings or unglazed pictures should be dusted carefully, and be professionally cleaned when needed. These should not be hung where they are liable to collect dirt nor be in such places as above open fires or other sources of heat

Electrical fittings – If within reach, dust. It is often the practice to have fittings washed or cleaned on the replacement of bulbs or tubes

WORK CARRIED OUT PERIODICALLY

Floors and corridors – Remove old wax; re-wax as necessary

Ceilings and walls – Dust or vacuum. If the surface is suitable, wash with a cloth wrung out in detergent and water, rinse, and dry

Carpets – Shampoo. Frequency depends on the degree of soiling, but removal of fixed dirt will prolong the life and appearance of a carpet

Upholstery – Shampoo yearly or as required

Curtaining – Heavy curtains washed or dry cleaned yearly; more frequent in areas with high air pollution or where there is heavy smoking by the residents. Net curtains as needed. The periodical vacuuming of curtains to remove dust will reduce the need for so frequent washing

Blankets and covers – Washed or dry cleaned yearly or as needed

Cupboards and drawers – Remove, empty, dust, wipe out, and re-line if surface requires it. In an hotel or where residents are constantly changing, drawers have to be checked and cleaned at each change, preferably before the guest departs so that any left property can be returned

Electric light shades and fittings – After first switching off the electrical supply, clean both sides of shades, bulbs, and tubes, and any securing flexes or chains

In large-scale units all of these housekeeping procedures may not be under the control of the Housekeeper; public rooms, entrances, and reception areas are often cleaned by day or night portering staff under the control of the Head Porter, whilst much of the periodic cleaning may be done by the Maintenance staff. This means that there must be full co-operation between all departments.

To ensure that regular cleaning procedures are carried out Cleaning and Maintenance cards may be kept for each section to record cleaning schedules; these are usually operated on a yearly basis or on a 12-week cycle. An example is shown in Figure 6.3.

All rooms which are occupied or in use are cleaned daily; an unoccupied room, once cleaned, is maintained by a quick dust and airing each week. An unoccupied bedroom would require an allowance of about 15 minutes of cleaning time each week, whilst one which is occupied will have an allowance of about 2 to 2½ hours each week, depending on its size, type of furniture, and the agreed cleaning frequency. In a 4-star hotel this time allowance might be doubled (*see* Chapter 9).

Weekly cleaning processes are added to the daily cleaning by either:

(*a*) Carrying out all the processes in a room on the same day, so that three or four rooms are daily/weekly cleaned each day and the remainder are daily cleaned, or

(*b*) Adding one of the weekly cleaning jobs to the daily routine, so that by the end of the week all rooms have been correctly cleaned. The routine might be:

Daily Cleaning Routine plus:

Monday – Paintwork, doors, and windows

Tuesday – Upholstery and wall dusting

Wednesday – Furniture polished and/or bed-making

Thursday – Floors and corridors

Friday – Windows and mirrors

The order of work would be designed to fit in with the needs of the establishment. In most places daily cleaning is usually all that is carried out at week-ends as this is the time when fewer hours are worked by the staff as days off are frequently taken.

Each room may be daily or daily/weekly cleaned individually or, where there is a block of similar rooms, it may be found quicker to use the *mass production* method wherein each process is carried out in rotation throughout the block of rooms; that is, they are all dusted in turn, vacuumed, etc. . . . In this way, the maid uses one set of equipment and materials at a time and does not have to keep changing her tools.

Whichever way of cleaning is adopted, the maid must be taught to

Planned cleaning and maintenance tasks						
Section, floor, department, etc.						
Week	Date	Windows	Curtains	Floors	Lights	Remarks

FIGURE 6.3 Cleaning and maintenance cards

be flexible and to fit in with the convenience of guests or residents; although, if alterations of routine are too frequent, it may be necessary for the Housekeeper to adjust work times and schedules.

METHODS OF WORK

All cleaning is governed by the following *basic organizational principles:*

(*a*) Any process which creates dust is done first. This allows time for the dust, which becomes air-borne, to re-settle and be removed during the final dusting. Beds, in particular, are made early in the room-cleaning routine.

(*b*) Whatever process is being carried out, it should not impinge on or damage any other surface which should be protected where necessary. A common example of damage caused is the placing of a bucket containing hot detergent and water on a polished floor or a carpet.

This will remove polish and leave an unsightly ring mark; or, if water is spilt on the carpet may cause colour running or bleaching. A rubber mat or paper placed under the bucket will protect the surrounding surfaces.

(*c*) All dirty or wet processes are carried out before final vacuuming, furniture dusting, and polishing.

(*d*) Polishing is left to be done as late as possible in the routine to avoid spoiling the surfaces by further handling.

(*e*) Care must be taken with all equipment and materials so that they are not placed where they block corridors or other traffic lanes with the possibility of causing accidents or damage. An example of this is placing opened tins of polish where they can be knocked over. Safe methods of working must always be encouraged.

(*f*) Whether it is part of the daily cleaning routine or not, all stains and marks *must* be removed or reported as soon as possible so that they can be taken out before they become fixed and permanent. This cannot be stressed too much to staff.

A SAMPLE METHOD OF WORK

A sample method of work for cleaning a study-bedroom with wash-handbasin, partly fitted furniture, rugs, and a polished vinyl floor would be:

For daily and daily/weekly cleaning

1. Open windows to ventilate room.
2. Remove waste-paper baskets, ash-trays, and other debris; empty and clean.
3. Clean rugs and roll up. Place outside or out of the way.
4. Make bed (*see* Figure 6.4).
5. Clean wash-basin.
6. Sweep floor, using impregnated mop.
7. Dust all horizontal surfaces; start at door and work methodically around the room, not forgetting any objects in the centre.
8. Replace rugs, waste-paper baskets, and ash-trays.
9. Close windows. Make sure all is clean and tidy.
10. Lock room.

[*contd. on page* 77]

Notes for Instructing Staff

HOW TO MAKE AND TURN DOWN A BED

What to do	*Main points to mention*
Collect clean linen	
Put chair(s) at end of bed	Remember to air room
Loosen blankets and sheets; place pillow on chair	
Remove blankets and place on chair	Stand at side of bed and fold in three, place the top edge to middle, then bottom edge to middle; this saves time both in removing and replacing and bed-making becomes automatic
Remove sheets	Fold in the same way and place on one side for the linen-room
Turn and air mattress	Explain that this is dependent on the type of mattress. State how often an interior-sprung mattress has to be aired and turned
Replace or straighten underblanket	
Put on lower sheet and mitre or envelope the corners	Sheet placed with fold in centre of bed, right side up. Explain that mitred or enveloped corners help to give a smooth appearance to the bed and retain the bedding firmly in position
Put on top sheet and mitre the bottom corners. Tuck in at bottom but leave sides loose	Place sheet level with the top of bed; this gives an extra 'tuck-in' at the bottom and does not leave too much sheet to turn over the blanket
Replace blankets. Mitre corners. Turn top sheet over blankets and tuck in sheet and blankets at sides.	Explain the procedure for turning the bed down at night, when blankets and top sheet are turned back to make a neat diagonal fold with the bottom sheet left smooth and well tucked in. Night attire is placed neatly by the turned back fold ready for use
Put clean cover on pillow, shake, and replace	The bed should look neat and tidy and the sheets be uncreased
Replace eiderdown and quilt	Show how to remove quilt at night and where it is to be placed

FIGURE 6.4

Jobs on the weekly cleaning list are inserted in the appropriate place. Furniture polishing will be completed between tasks 7 and 8, window cleaning between tasks 5 and 6, and the floor will be polished after task 7.

POINTS TO BE STRESSED

Points to be stressed are:

(*a*) The safeguarding of a resident's property. In a number of establishments the maids are instructed to report to the Housekeeper when valuables or money are left in the rooms so that she may advise the owner on their safer disposal.

(*b*) Room maintenance requirements – such as loose door knobs or broken sockets – must be reported.

(*c*) The care and safe-keeping of keys and master keys; and the necessity for locking rooms after cleaning and, if a room is on the ground floor, for closing or locking windows.

(*d*) The handing-in of all lost property.

ROOM INSPECTION CHECK LIST

Inspection is important to maintain standards and to ensure a continuing good room appearance. A Housekeeper may decide to inspect all rooms daily or be content to make a sample inspection of three or four rooms daily in each section. She may even only check informally as she goes about the building or when she has occasion to visit the maid on each floor. A surprising amount can be seen by just standing at an opened door: are the curtains arranged tidily, are the covers straight, is there any dust or marks on the floor, are the table tops clean with no ring marks from the morning cup of tea? Does the room smell fresh with no remains of tobacco smoke?

A more thorough check can reveal dust behind doors, wash-handbasin grilles with hair, or a layer of powdered abrasive not rinsed from the bath left ready for the next occupant.

Experience teaches the main points to look for. These are:

Door

Lock, key, hinges; do they need oiling?

Windows

Ease of opening and closing; cleanliness of glass; condition of window locks; curtains arranged tidily and all hooks and end-stops in position; blinds, shutters, or venetian blinds in good working order.

Electrical fittings

Switches and sockets, light bulbs, service bells, light pull cords in bathrooms and above beds, razor sockets.

Telephone, Radio, and Television

Plumbing

Rate of flow of water from bath and wash-handbasins; taps for stiffness or drips; flow of water from cistern to W.C.; the condition of chrome fittings and mirrors.

Heating

Efficiency; ease of control; ventilation control, if air-conditioned.

Plus

General cleanliness and appearance, and comfort of beds and other furniture.

PERIODICAL, SPECIAL, OR VACATIONAL CLEANING

In most establishments, this has to be planned for and fitted into slack periods and completed either by the closing of a floor or wing, by the thorough cleaning of individual rooms as they fall vacant for two or three days, or the cleaning is arranged to coincide with the planned schedule for redecoration.

Because of the difficulty of maintaining a flexible programme for periodic cleaning, records must be kept so that it can be clearly seen when each sector was last dealt with and when the special cleaning is again due. Establishments which close for stated periods during the year for Christmas, Easter, or the summer vacation, may still have trouble keeping to schedule as this is the time when staff vanish on holiday or revert to a part-time retainer basis of working.

The main advantages of planned routines for all cleaning are:

(*a*) It does ensure that the entire building *is* cleaned and some parts are not forgotten.

(*b*) The work is evenly allocated.

(*c*) The work is completed at the right times.

(*d*) The housekeeper does know who is responsible for each section.

(*e*) The right equipment and materials can be made available.

(*f*) If other deparments are involved, teamwork can be co-ordinated.

(*g*) The working of the whole department is clearly defined; a new-comer should be able to take over the running of the department without upsetting existing routines.

COMPARISON OF CLEANING COSTS

It is very difficult to compare costs of cleaning of one establishment with another as conditions and service requirements can vary tremendously; but for a building of modern design and construction it would be expected that costs per unit area should be within a range of + or − 20% of the average costs for similar buildings. To find the cost:

The total costs for labour, equipment, and materials is divided by the area cleaned to give the cost per m² (ft²). For an example: A modern building which is mechanized and has an area of 23 225m² (250 000 ft²).

1. LABOUR COSTS

400 hours per week @ 25p = per year	–		5 200·00
Supervisor @ £720 per year	–		720·00

2. EQUIPMENT

Tools, cloths, etc. per year	–		100·00
Capital equipment: 8 Suction cleaners @ £75 each	–	£600·00	
1 Scrubber @ £120	–	£120·00	
3 Polishers @ £85	–	£255·00	
1 Dustette @ £15	–	£15·00	
		£990·00	
Cost of capital equipment depreciated over 10 years	–		98·00

3. CLEANING MATERIALS

Approximate cost per year	–		250·00
Total Cost (1 + 2 + 3)	–		6 369·00

The cost per m² per year will be divided by $\frac{£6\,368}{23\,225\ m^2}$

which is 2½p per m²

It is only by producing and comparing figures such as these, that one can tell if the cleaning and housekeeping department is really operating efficiently.

In a survey on inter-hotel comparisons, undertaken by the University of Surrey, cleaning costs for small to medium hotels averaged from £27·08 to £35 per 1 000 ft².

ELECTRICAL CLEANING EQUIPMENT

Detailed figures given by Work Study experts are as follows. But these figures may be quite different for other circumstances and equipment.

To buff a floor of 100 ft^2 (9·29 m^2)
= 6·44 min using a 'bumper' or weighted polishing mop
To buff a floor of 100 ft^2 (9·29 m^2)
= 3·11 min using a floor polisher
To scrub a wooden floor of the same area
= 14·6 min by hand
To scrub a wooden floor of the same area
= 6·9 min using a combined scrubbing/pick-up machine. To this time must, however, be added the time for filling, emptying, cleaning, and putting away the equipment; nevertheless, for large areas the overall saving in labour is considerable.

It is because of comparison timings, such as the above, that more and more firms are investing in large-scale industrial equipment. Machines are available which will scrub, strip, scour, polish, wet and dry vacuum, and shampoo using either a liquid or dry foam. Some machines are on the market which will combine up to 5 of these functions. The choice of machinery depends on:

1. The type of work required, the area, and the cleaning frequency.
2. The size and loading (wattage) of the equipment suitable to the area to be covered.
3. The initial cost of the machine and the estimated cost of working.
4. Availability of regular servicing and maintenance. For the convenience of servicing it may be policy for all equipment to be bought from the same manufacturer.
5. Its safety in operation, both mechanically and electrically.
6. The ease of operation and adaptation for different attachment fittings.
7. Operating noise.
8. Its mobility and stability; it must not be top-heavy.
9. Weight and handling; can women cleaners operate the machine easily?
10. Good rubber or plastic buffers fitted so that damage will not be caused to skirting-boards, doors, and furniture.

Domestic sizes should be avoided as the machines are too light and, as the rating is lower than in the commercial models, the motors tend to burn out more quickly with heavy use.

Most equipment carries either a one- or two-year guarantee and will have a 'write-off' period of from 5 to 10 years depending on the policy of the firm and the estimated amount of use that it will have.

In addition to the above points the main requirements for equipment are as follows.

DRY VACUUM OR SUCTION CLEANERS

The choice is between the upright and the cylindrical type of cleaner. Both use suction power but the upright model has, in addition, brushes which are rotated by the motor and remove dust by beating. These are most suitable for carpeted rooms, where the beater bar action is very effective in removing both surface and embedded dirt and grit. When buying, the main points to watch are:

(*a*) Good suction power. This is related to the loading (wattage) of the motor.

(*b*) Ease of emptying with the dust going into either disposable paper bags or a metal container in the cylinder. This should be light and readily emptied.

(*c*) On all cylinder cleaners the return air is filtered before release back into the atmosphere. This filter must be efficient to prevent fine particles of dust or bacteria being blown back into the room, particularly where hospital cleaning is concerned. Where an air-return filter is provided, this should be cleaned regularly.

(*d*) The attachments. With both types, attachments are provided for high dusting, for upholstery and curtain cleaning, and for removing dust from crevices and the edges of fitted carpets. These tools must be easily interchangeable but should also fit firmly so that they cannot become detached in use. Flexible hoses must be well balanced and of medium weight as, if too heavy, they are difficult to handle and liable to knock against furniture. Swivel joints are an advantage so that cleaning can be done at any angle.

All dust bags and cylinders must be emptied regularly and not allowed to become over-full as this puts extra strain on the motor and will prevent pick-up.

Some establishments, notably the newer hospitals and hotels, have built-in suction installations; these have a powerful motor, fan, and dust container with tubing running through ducts in the walls which connect the power unit to the outlet points in the skirtings of the room. Flexible hose is connected for cleaning. The main advantages of this method are that motor noise is greatly reduced with the only sound coming from the dust entering the pick-up tool, and dust is centrally collected so that the emptying of individual machines is eliminated.

DUSTETTES

These are the hand-operated suction cleaners of use for stair-cleaning and for upholstery.

WET VACUUM PICK-UP

The additional requirements are for:

(*a*) Rust-proofed tanks. Stainless steel are best for long life.

(*b*) Size. These machines are heavy when full and so difficult to use by women cleaners or over uneven floor surfaces. Machines should not be bigger than a 22 or 27 litres (5 or 6 gallons) capacity.

(*c*) Emptying. The best machines have a tap fitting or hose which is connected to the base so that the tank can be emptied over a floor drain. Others require tilting and lifting.

(*d*) The type of pick-up. This can be either a fixed squeegee head or else a flexible suction hose.

These machines must always be emptied after use, dried, and stored in an open position so that they are well ventilated.

POLISHING/SCRUBBING MACHINES

If the machine is to be used solely for polishing, then a combined suction-polisher is often bought. The suction nozzle placed in front of the brushes removes dust from the floor before polishing takes place and prevents dust being scattered. These dual-purpose machines only remove light dust: heavy debris should be removed before the machine is used.

If the machine is to be used for polishing, scrubbing, stripping, scouring, or carpet shampooing three grades are available (for light, medium or heavy duty). Their main requirements are:

(*a*) Manœuvrability into corners and under pipes and radiators and the ability to clean close to the skirting boards.

(*b*) Ease of filling and emptying the solution tank and easy control of water/detergent solution to the brushes for scrubbing.

(*c*) Type of brush head, either single or multiple. If there is a single head it is usually possible to reverse the rotation direction of the brush to even out the wear. Multiple brush heads can be sprung to allow an independent up-and-down movement for uneven floors.

(*d*) The machine should be of an appropriate size for the area concerned as the advantages of a large machine of this type are lost through poor manœuvrability when ed in small areas.

(*e*) The size of brush head.

A 12-inch brush cleans up to 92·9 m^2 (1 000 ft^2) per hour

A 14-inch brush cleans from 111·48 to 185·8 m^2 (1 200 to 2 000 ft^2) per hour

A 16-inch brush cleans from 185·8 to 371·6 m^2 (2 000 to 4 000 ft^2) per hour

A 19-inch brush cleans from 371·6 to 743·2 m^2 (4 000 to 8 000 ft^2 per) hour

These timings are inter-dependent on the condition of the floor surface and on the loading of the machine.

High-speed machines are now being brought into general use. These have a brush speed of between 300 and 400 revolutions per minute but, with present design, can only be used for spray cleaning and polishing and are unsuitable for scrubbing or carpet shampooing.

The comparative costs of the different sizes of machine are shown by the following calculations:

	A machine with a 12 inch brush head	*A machine with an 18 inch brush head*
Cost + attachments	£45 + £12	£120 + £25
Area to be cleaned – 6 days per week	743·2 in^2 (8 000 ft^2)	743·2 m^2 (8 000 ft^2)
Time required per 92·9 m^2 (1 000 ft^2)	40 min	20 min
Total hours per day	5⅓ hours	2⅔ hours
Labour costs per hour	55½ p	55½ p
Cost per year	£956·80	£478·40
Total cost per year	£1 013.80	£623.40
or with labour costs at 25p per hour:		
	£416 + £57	£208 + £145
giving a total cost	£473	£353

The yearly savings are very apparent.

Attachments are polishing and scrubbing heads and a wide range of metal, fibre, and nylon wet-pads for scouring and stripping.

Working on the same principles as the above are the carpet shampooers and the combination machines used for scrubbing, drying, or wet pick-up.

All new equipment should be demonstrated on site and used by the cleaners concerned so that the ease of handling can be gauged.

Most large-scale cleaning suppliers will give advice; some will train staff in the use of their equipment; and a few will prepare cleaning schedules in their search for satisfied customers.

As with other types of equipment, cleaners, carpet shampooers, and other electrical equipment can be rented at highly competitive terms.

OTHER CLEANING EQUIPMENT

	Use and Care	*Selection*
Carpet sweeper	To remove surface dust and crumbs from carpets. After use, empty and wipe over with damp cloth. Wash brushes periodically. Keep wheels lubricated	
Mop buckets	After use, empty, clean, rinse, and dry to prevent development of rust and smells	Consider those with wringer attachments for greater efficiency. Buy to suit size of mop used
Yarn (wet) mops	Rinse well or wash after use. Hang up to dry so that heads do not rest on floor	Consider size, up to 24 oz. A high percentage of jute to linen/cotton in the mop will not wear as well as all linen or cotton
Impregnated (dry) mops	Never use on wet floors. Collect dust and clean after use, using vacuum to remove dirt if possible; clean daily. When dirty, wash, rinse well and, when dry, spray with impregnated solution	Correct size. Ease of fitting to mop heads and attachments
Weighted polishing brush ('bumper')	The weighted brush head has a flat surface which is covered by a clean duster or cloth	These vary in size and weight. They are not now in general use but useful for quick manual polishing
Squeegees	Clean and wipe after use	Size to suit cleaning area. Ease of fitting
Spray	For application of polish. Clean regularly to prevent clogging of jet. Empty after use	Size 1¾ to 2 pints (1·14 litres) for general use. Container which will not corrode with the polish
Applicators	For the application of polish. Made of sponge or lambswool. Must be washed after use	Ease of replacement of fitting
Maid's trolleys	Cleanliness. Lubrication of wheels	Consider design and construction: not too top-heavy when loaded; easily manœuvrable; large enough for equipment used but also correct size to fit cleaner's storage cupboard or area.

	Use and Care	*Selection*
		Storage attachment for rubbish or dirty linen. Wheels large enough to move easily over different floor surfaces
Step-ladder	Inspect for safety. Keep clean	Of light-weight metal. Platform-type to take cleaner's equipment
Dustbins (if used)	Empty daily, clean	Use polythene to prevent noise and denting, and for lightness. Size suitable for the average collection of rubbish
Paper sacks	Staple top over when full. Make sure top attachments fit correctly	Correct size and ply suitable for type of rubbish
Sanibins	Clean regularly	Polythene to prevent rusting, noise, and denting
Scrim	Always use damp. When dirty, rinse out in plain warm water and hang to dry. Used for polishing glass, mirrors, etc.	Ordinary scrim is 40% jute and will last about 2 weeks with heavy use. Linen scrim will last 6 weeks or more with the same type of use
Mutton cloth, net cloths, floor cloths	Rinse well after use and hang to dry	Must be absorbent and correct size and weight for the cleaner's use
Chamois leather	Never use with hot water or dry over a radiator or other hot place; this hardens and splits the leather. Do not use with acids or any solvent or with an alkali detergent. Rinse in warm water after use; when dirty wash in good quality soap flakes; there is no need to rinse the soap out	Buy only good quality as cheap leathers split and fall to pieces after a little use. Quality is not indicated by thickness but by the amount of stretch there is in the skin; a better quality has more stretch than a poorer leather
Sponges	Always rinse after use; if very dirty, boil for a few minutes with a little soda. Squeeze out water, never wring out; dry well before storing. Do not use on roughened surfaces	Synthetic sponges available in different sizes depending on use. Must be highly absorbent
Dusters	Wash after use. Use folded to collect dust and not to scatter it	Consider use of impregnated dusters or gloves

CLEANING MATERIALS

FLOOR SEALS

Any floor surface which is porous and has a granular texture requires to be protected by a seal. This is applied to prevent dirt and grease becoming embedded in the surface. Wood, wood composition, and cork floors must always be sealed. Magnesite, flexible P.V.C., concrete, asphalt, and linoleum, can be sealed if situated in heavy traffic areas and where maintenance is difficult.

A seal is a semi-permanent liquid material which penetrates into the surface of the floor filling the open pores and, when dry, hardens to give a smooth, durable surface which will protect it and, in so doing, extend the life and appearance.

Seals are semi-permanent. When wood, wood composition, or cork floors are laid, the seal is usually applied directly by the contractors. Linoleum can be bought sealed or un-sealed; other floors where the seal is optional may be treated by the Housekeeping Department. The life of a good seal depends on good preparation of the floor before application, the amount of wear and use and, to a certain extent, the cleaning treatment. Protection can be given by a wax polish finish. An uneven sub-floor will cause the seal to crack.

TYPES OF SEAL

(a) Oleo-resin

These are easy to use and to apply. They consist of oil processed with resin, with the addition of thinners and dryers. These seals are very suitable for light traffic areas but, with heavy use, worn patches will appear which can however be readily touched up with the seal as they occur. The seal normally hardens in 8 to 10 hours but, as with other seals, the longer the floor is left the more durable the surface becomes. A slight disadvantage is that the oleo-resin tends to darken the floor; this is particularly noticeable with a light porous surface such as a honey-coloured cork.

(b) One-pot plastic

These seals contain no drying oil and harden with the evaporation of a solvent or by the addition of a chemical hardener. These, again, are easy to apply and will dry in from 3 to 4 hours. Touching up is not so easy as with an oleo-resin seal and it is found often to be better to remove the old seal entirely and re-seal the whole floor when the surface becomes worn. This type of seal does not usually alter colour and is most suitable on light woods and corks.

(c) Two-pot plastic

These seals require more care as two materials, a base and a hardener, are blended together before use. This means that care is necessary to see that the right proportions are used and also to ensure that the mixture is not left for any length of time before work starts. Hardening starts immediately so that any delay in application either results in the seal having insufficient time to penetrate or it will thicken and become unusable. This type of seal gives a much harder finish and a better gloss, and is more resistant to chemical and abrasive damage than types (*a*) and (*b*), but touching-up is difficult and is not really satisfactory; the best treatment is to strip and re-apply.

(d) Pigmented seals

When the colour of the floor surface, usually a wood or concrete floor, needs brightening or altering, a pigmented seal is often used; these can be either one or two-pot seals.

Good preparation of the floor surface is essential before a seal is applied. The floor must be completely clean and free from all old polish, seal, dirt, and grease. Wood or cork floors are sanded to remove all unevenness before treatment and to provide a roughened surface so that the seal will key or bond to the floor. The seal is best applied at room temperatures with good ventilation to ensure a quick drying time, and the room should be kept as dust-free as possible during this period.

The life of a seal varies but can be extended to two or three years if the correct maintenance has been applied; many seals are protected by regular applications of polish.

The most well-known study on floor maintenance and treatment was carried out at the Ashford Hospital in Middlesex in association with the King Edward's Hospital Fund for London. In their report they compare the treatment and durability of ten different floor surfaces: P.V.C., cork, linoleum, rubber, vinyl asbestos, etc., the life of various seals, the initial cost of the flooring, and the estimated annual labour costs for maintenance for each type. The experiment was carried out in a main corridor where there is heavy traffic and covered a period of two to three years.

The conventional solvent-based seal should not be used on thermoplastic, rubber, vinyl asbestos, or terrazzo floorings. It is not always considered necessary to seal these but if it is required a water-based seal should be used.

Seal is removed by the application of a stripping agent or approved solvent which is supplied by the manufacturer of the seal and is adapted to its special properties. All old wax must be removed and the floor roughened to provide a key for the new seal by using a coarse grade wire wool or nylon web scrubbing pad.

FLOOR WAXES OR POLISHES

Until 1929 the only type of wax in use was solvent-based; it consisted of a suspension of wax in a solvent, usually a white spirit or paraffin: of these, white spirit is fast drying whilst the cheaper paraffin is slower drying and is used as a basis for the cheaper polishes. With the introduction of new types of floors, asphalt, rubber, and the later vinyl asbestos, it was found that the solvent-based waxes softened and gradually began to dissolve and crack the surface. This is not apparent after one or two applications but certainly is after several months of wrong treatment. This led to the introduction of the water-based waxes which are now in general use.

Polish must be applied thinly to give good results with two or three thin coats, which are buffed between applications, giving the most hard-wearing surface. If polish is applied too thickly, it is difficult and time-consuming to buff up and it will lead to a build-up of dirty polish on the floor.

When applying polish, very little should be used at wall and corridor edgings, where there is little wear; to prevent this build up of polish and dirt the polish should be applied only where needed on the main traffic areas.

Polish should not be re-applied too often. A buffable polish is easily restored to its original condition by buffing with a polishing machine. Rooms with little heavy traffic may only need a fresh application of polish every four to six months.

Solvent-based Waxes

Are used for wood, wood composition, cork, and linoleum floorings and are obtainable in paste or liquid form. Liquid waxes are an efficient cleaning agent but a paste wax, because of its high wax content, will not clean a floor in addition to polishing it.

Water-based Waxes

Can be used on all types of floors and are an emulsion obtained by heating the wax with an alkali, such as soda, and water. Much harder, higher melting waxes can be used than with a solvent-base so that the resulting polish is more durable. Shellac or resin is added to the wax to aid the hardening process and also to assist the spreading properties. In 1952, polymers were added. These are either styrene, which led to the dry-bright range of polish, or acrylic which are more durable but do not give such a high polish. A combination of both polymers can also be used. The three main constituents of a water-based polish are now:

Resin – to harden and to act as a levelling agent. Since it is soluble in alkali, an alkali detergent (or stripping agent) is used to remove old polish.

Wax – for buffing up to a polish.

A polymer – for durability.

The *proportions* of resin, wax, and polymer used depend on the use and properties required in the polish. A *high polymer content* will give a dry-bright finish but one which will not re-buff; so to renew the surface, fresh applications have to be made. It is not, therefore, suitable for any area where there is much hard wear, traffic, or dirt, but is used in rooms which have little use or wear. The great advantage is that, as it drys bright, no polishing is required.

A *high wax content* will give a matt finish but one which can be re-buffed when needed so that the frequent re-application of polish is unnecessary. Wearing qualities are good and both this and dry-bright polishes have good carbon-black heel mark, stain, and water resistance.

Wash and wax emulsions

Wash and wax emulsions are usually a dry-bright polish to which a detergent has been added so that the two components are stable together. Once a good surface has been obtained, the floor is cleaned and polished in one operation. As the 'wash and wax' is applied, the detergent loosens the dirt which is removed by the applicator, whilst at the same time a new layer of polish is applied. This gives good results on light-coloured surfaces as no build-up of wax or yellowing can occur as each coat of polish is partially removed by the following application.

Detergent-resistant emulsions

Detergent-resistant emulsions are built up to combine dry-bright and rebuffable qualities with resistance to detergent. This means that a floor can be cleaned and light dirt deposits removed with no damage to the film of floor wax. They are exceptionally durable even in heavy traffic areas and as such are, of course, difficult to remove. The polymer is combined with an acidic group bonded to a metal – either zinc or zirconium – and will not come off with conventional detergents. Removal is effected with a strong alkali or with ammonia which combines with the metal. Anti-slip properties can also be built into all polishes.

When an alkali detergent is used to remove old wax, it must be thoroughly rinsed from the floor with clear water or a neutralizer (an acid, such as vinegar). If this is not done, the alkali will continue its action on the new polish and will result in a patchy appearance.

All buffable waxes must be left to dry and harden before the floor is polished. The exception to this is when liquid solvent or water-based polishes are used for spray cleaning. For this, the waxes are sprayed on to the floor, which is polished with metal fibre or nylon web pads fitted to the polishing machine. This has the effect of cleaning and polishing at the same time with the dirt lifted from the floor and held in the pads. The pads should be replaced as they become dirty and then washed and re-used. This process leaves the floor in good condition and, if required, it can be buffed up normally using the polishing brushes on the machine.

CLEANING AGENTS

These are used for the removal of dirt which has become fixed or attached to a fabric or surface. Very light soiling is usually removed by wiping the surface with a cloth wrung out in water, but heavier soiling requires the use of an agent.

DETERGENTS

Detergents are defined as cleaning agents and have all or some of the following properties:

(*a*) To act as a water softener so preventing any deposit of scum or film of insoluble salts on the surface of the material being cleaned.

(*b*) To 'wet' the surface thoroughly so that the detergent is able to penetrate the fibres and surround all the dirt particles, so that the cleaning is uniform.

(*c*) To soften and loosen the dirt.

(*d*) To hold the dirt in suspension.

(*e*) To be effective in all temperatures of water.

(*f*) They must not affect the materials on which they are used nor must they have an adverse effect on the skin or the hands of the cleaners.

Detergents are only really efficient when used with some type of mechanical action such as occurs in mopping and scrubbing. Heavy dirt removal may also require the use of elbow grease.

Detergents are manufactured for various purposes and to suit the hardness of the water used with them. Strictly speaking, soap is a detergent but is not referred to as such in common speech.

SOAP

Soap is made from a combination of natural fats or fatty acids with an alkali and is not now in general use for cleaning because, when used in cold water or in hard water areas, it forms a scum and will leave a film on a surface. Foaming properties are also reduced with hard water so that dirt is not held in suspension as well as with the synthetic detergents.

SYNTHETIC DETERGENTS

These were first made in 1913, but not brought into general use until 1939. For normal cleaning purposes detergents are divided into two groups: neutral and alkali, depending on the pH value.

The pH scale

The pH scale ranges from 0 to 14. 0 to 6 is acid, 8 to 14 is alkali, and the pH value of 7 is neutral, but as the degree of acidity or alkalinity increases gradually from neutral, the pH values 7 to 10 are usually considered to be neutral by most manufacturers.

0 1 2 3 4 5 6 7 8 9 10 11 12 13 14
←——acid——————neutral——————alkali——————→

For most cleaning needs a neutral detergent is used as this does little damage to surfaces or to the cleaner's hands but, as the pH value of a detergent increases, it becomes more effective as a dirt remover and also becomes more destructive. Alkali detergents are only used where there is a heavy build-up of dirt, and care must be taken to rinse the surface well after use. When soap is used, the dirt is suspended in the foam and bubbles produced; this property is not necessary in a detergent but is normally added because most cleaners expect a cleaning agent to produce foam and lather and will continually add more detergent to water until a foam is produced. This foaming property prevents waste. When buying, the dilution rate and working strength should be checked; what may seem a very good buy may not be as economical as a more expensive product. Dilution rates should also be given to the cleaners who are using the detergent.

When detergents are used constantly by cleaners and washers-up, grease is removed from the skin. This causes dryness and, in extreme cases, dermatitis. To prevent this happening, a personal issue of rubber gloves is often made.

Detergents are also used for hand-washing and in dispensers in public cloakrooms but in most other places soap tablets are used. These can be bought in ¾, 1½, 2, or 3 ounce size depending on the usage and the length of stay of a guest. (The metric equivalents would be 18, 35, 45, or 68 ml).

ABRASIVES

Abrasives are used when the dirt is embedded or fixed particularly firmly to the surface and are graded to suit the material on which they are used.

All abrasives should be used only on the materials for which they have been designed as considerable damage can be caused by using the wrong grade.

Types of abrasive	*To clean*
Very fine: pink oxide of iron or jeweller's rouge	Silver and gold
Slightly coarser: precipitated whiting	Brass and harder metals
Powdered pumice, in paste form	Baths and other bathroom fittings
A coarser grade pumice	For heavier kitchen use
Nylon web pads and metal fibre pads	For floor cleaning and stripping
Coarse sanding pads	For renewing wood and floor surfaces

Depending on the properties required, abrasives can be mixed with soap, synthetic detergent, alkali, bleach, glycerine, or a solvent. When using any abrasive, care should be taken to rinse it off again, otherwise the surface will be left smeared and rough to the touch.

METAL POLISHES

Metals are polished for two reasons:

(*a*) To remove tarnish,

(*b*) To remove any scratches on the metal caused by use, washing, and general handling.

Tarnish can be removed by either the mechanical abrasive action of the polish or by a chemical reaction obtained by immersing the metal in a solution which reacts with the sulphides in the tarnish and frees them from the surface. Scratches are 'healed' by the flowing of the surface metal into the scratch; this is achieved by rubbing the metal with a fine abrasive in the presence of a fatty acid. Burnishing machines do this by the rotation of highly polished steel balls which are immersed in a detergent solution. The balls, rolling against the surface of the metal, burnish and remove tarnish.

Silver dips are essentially chemical cleaners and will not remove any scratch marks from the surface.

Some proprietary makes of metal polish now contain a 'long-life' element. This forms a film which combines with the metal; it is colourless and does not affect the appearance of the article but does help to prevent tarnishing. This protection is affected by some acids or alkalis but, with care, a bright shiny surface can be maintained for several months (*see* Chapter 3).

Metals fall into two groups: soft metals which are silver, gold, and some of the pewter alloys, and hard metals which are brass, copper, zinc, and the majority of the pewter alloys.

A polish for soft metals contains jeweller's rouge or very fine powdered whiting and a fatty acid suspended in a water/alcohol or petroleum-based medium. For hard metals, a coarser abrasive is used. To prevent scratching, soft polishing cloths should always be used and, if the object is to be used for food service or preparation, it must always be washed before use.

The metals which require regular attention are silver, E.P.N.S., brass, and copper all of which react with the oxygen and sulphur in the air to form silver sulphides, copper carbonates, or verdigris. Brass or copper which is exposed needs constant care. When used for door-plates or guards, these can be treated with a transparent lacquer which will prevent tarnish developing. The lacquer does, however, slightly alter the appearance of the metal and should never be used on any object of value.

Stainless steel requires a special proprietary cleaner which will not react chemically with the surface.

STAIN REMOVAL FROM METALS

Silver, rub with salt, then wash.
Copper, rub with lemon or vinegar and salt.

DISINFECTANTS

In the Housekeeping Department, disinfectants are used in routine cleaning to control infection and the growth of any pathogenic organism which may be present. Disinfectants are used in toilets and bathrooms, sick quarters, occasionally as a sterilizing agent in crockery and cutlery washing and, by the House Porters, to cleanse all outside drains. Because some organisms become resistant to a particular disinfectant, the effectiveness for hospital use should be tested regularly under local working conditions.

The efficiency of a disinfectant is measured by the phenol coefficient introduced by Rideal and Walker (R.W. number) in which all disinfectants and antiseptics are compared with phenol which is used as a standard comparison.

The dilution rate is found by the following formula:

$$\text{Dilution rate} = \frac{1}{20 \times \text{phenol coefficient}}$$

As an example: if phenol concentration is R.W. 4, the dilution rate is

$$\frac{1}{20 \times 4} = \frac{1}{80}$$

or one unit of disinfectant to 80 units of water.

To be effective:

(*a*) Disinfectant must come in direct contact with the organism; therefore any layers of dirt or grease must first be removed.

(*b*) It must remain in contact with the organism sufficiently long to kill; the time may be affected by the temperature at which it is used.

(*c*) Moisture must be present.

(*d*) The disinfectant must not be neutralized by other substances; some of the common substances which affect its efficiency are the presence of some proteins, soaps, some types of detergents, and also the degree of hardness of the water which may either reduce efficiency or alter the dilution rates.

(*e*) Disinfectants have varying effects on different types of organism; in normal use this is unimportant, but for specialized hospital use a disinfectant may be needed to suit particular requirements.

When buying, consideration should be given, as with detergents, to the effective dilution rate and also to its stability in solution so that its effectiveness does not decrease with storage.

OTHER CLEANING AGENTS

Furniture polishes

Furniture polishes are blends of waxes, a spirit solvent, and often a silicone which makes the polish easier to apply and gives added resistance to marking from moisture or heat. As with floor polishes, the polish used on furniture should be applied sparingly with a soft cloth and rubbed up well, preferably by using a dry cloth in each hand to prevent finger marks being left on the surface of the wood.

Polishes can be obtained in wax, liquid or cream, or in aerosol sprays.

Teak is cleaned with a special teak oil or cream which leaves the wood with a matt finish.

Window cleaners

Window cleaners are made from a water-miscible solvent to which a small amount of synthetic detergent or alkali has been added and, occasionally, a fine abrasive to increase its cleaning effect. As water, water/vinegar, or water and methylated spirits, are equally effective and cheaper to use, although requiring a little more rubbing and elbow grease, a considerable number of establishments do not use proprietary brands of window cleaners.

Toilet cleaners

Toilet cleaners are acid, often based on sodium acid sulphate in crystalline form, and as such should not be used on any surface other than that of the W.C. pan as they have a corrosive effect. To be efficient the cleaner is sprinkled on the sides of the pan and into the water and to be effective, should be left as long as possible.

Bleach is also used for cleaning W.C. pans but should never be used at the same time as an acid cleaner as together they combine to produce chlorine gas, which is poisonous and highly undesirable.

Soda

Soda is an alkali and is used as a cheap grease emulsifier both in the kitchen and by the House Porters for cleaning outside drains and pipes. Used with boiling water the drain or pipe is flushed completely to get maximum effect.

Vinegar

Vinegar is frequently used, diluted with water, for window cleaning, the removal of stickiness and light dirt from wood, and as a neutralizing agent. Where there is the danger of colour run in carpets or furnishings, vinegar added to the water should stabilize the dye.

Methylated spirit

Methylated spirit is very effective for window cleaning and for the cleaning of mirrors and pictures where the use of water might cause damage and marking to the picture or mirror backing. It is, however, very inflammable so care must be taken when it is used. It is highly effective in removing hair lacquer spray from a mirror surface.

Bleach

Bleach is frequently used as a cheap cleanser and disinfectant for W.C. pans, sinks, and drains, and as a method of removing tannin staining from the inside of metal tea-pots (but please rinse well after use). The use and dilution rate should be clearly stated as many cleaners use it as they would water and pour as they think fit. To be effective, 60 parts per million of available chlorine should be in the 2 litres ($\frac{1}{2}$ gal) of water remaining in the W.C. pan. To achieve this requires only 1·5 g (0·05 oz) of bleach – well under a dessertspoon.

CLEANER'S STORES AND CUPBOARDS

Storage for cleaning supplies should be central both for the delivery of equipment and materials (which are usually heavy) from the suppliers and for ease of distribution to the cleaners.

Storage space should be sufficient for bulk purchases of drums and cases, for the storage of large-scale equipment and replacement brushes, mop-heads, mats, light bulbs, and fluorescent tubes, and also to provide sufficient room for the break-down from bulk-to-bottle or tin distribution to the cleaners.

A very rough guide to the area required is 1% of the area maintained up to 464·5 m^2 (50 000 ft^2), decreasing to $\frac{1}{2}$% as the cleaning area increases. It also depends on the amount of stock to be kept.

To prevent waste of time, stores are usually distributed weekly on a full-for-empty bottle or tin basis or new for old. This prevents the accumulation of half-empty dried-up tins or bottles in a cleaner's cupboard and helps to control the usage.

Liquid and powdered cleaning agents, bought in bulk, should always be issued in clearly marked containers so that dilution rates are known and misuse is avoided.

Stores should have an in-the-floor mop-sink to facilitate emptying and filling buckets and containers.

Cleaner's cupboards should be provided on each floor and be so arranged that all equipment can be stored out of sight and tidily. For preference, these cupboards should also be fitted with an in-the-floor mop-sink with hot and cold water, shelves for small equipment, and space for hanging mops, dust-pans, and sweepers, and for the storage of the maid's trolley and electrical equipment.

Each cleaner is responsible for the cleanliness and tidiness of her own cupboard and time should be allowed for her to keep it in good order. A good Housekeeper would include the cleaner's equipment and cupboard in the daily inspection of each department.

7 STAFF AND MANAGEMENT

A GREAT deal has been written about the selection, training, and management of staff all of which implies that there is considerable choice in the matter. Unless staff are being recruited for an entirely new building, however, when it is possible to start from the beginning it is found on taking up a new position that one has inherited a mixed collection of willing workers – good, bad, and intermediate. It is left to the management to weld them into the co-operative team which will make the establishment the most efficient and most desired place of work in the neighbourhood.

When meeting existing staff it is best to meet them collectively, perhaps over the morning cup of coffee, and then to meet them individually at their work so that one is able to talk to each about the job and their interests. It helps if one is able to remember names and faces easily.

It is only after three or four weeks when there has been time to appreciate the various existing timetables and routines that one would gradually bring in any alterations which are felt to be applicable to the work involved. Any alterations should be explained and discussed with the employees concerned and their opinions sought; often they will have very good suggestions of their own which should be considered and, perhaps, put into practice.

As staff retire or move to more lucrative fields one is faced with their replacement. The choice is not always extensive.

RECRUITMENT

The most usual methods are as follows:

(*a*) By advertisement quoting such salient points as type of job, hours, rates of pay, qualifications, whom to apply to, where, and when.

(*b*) By public employment agencies and the local employment exchanges of the Department of Employment.

(*c*) By contact with local schools and colleges.

(*d*) By the time-honoured method of asking the staff 'do you know anyone who would take Annie's place?' Recommendations received in this way are good, as no self-respecting staff are going to suggest anyone who will not fit in or will be reluctant to do her fair share of the work: their honour is at stake. One may find, of course, that they are suggesting other people's staff which can lead to difficulties particularly if one engages them. There can also be difficulties in supervision if the proportion of relatives and close friends is high in a small establishment.

All, or the best of the applicants, are invited for an interview.

THE INTERVIEW

This is usually conducted by either the Head of Department or by a representative of Management or of the Personnel Department in conjunction with the Departmental Head. With two people interviewing, it means there are either two short interviews with each person, and notes and recommendations are compared afterwards, or the applicant is interviewed by both selectors together. In the latter case, it must be clearly agreed between them, before the interview starts, which questions are to be covered – who will talk about the firm's policy, who asks about background and past experience – so that a clear unconfused impression is given. Another requirement is a quiet office with prior arrangements made to guard against any untoward interruptions. The establishment may normally be chaotic but there is no point in letting the applicant know this too soon.

A large establishment may require the applicant to fill in an *Application Form* before the interview, so that all the relevant information can be quickly studied. The use of a form may depend on the grade of staff being interviewed; many employees unused to clerical work find them discouraging and may need help in filling them in. There are two main purposes in an interview:

(*a*) To allow the interviewer time to know and assess the capabilities of the applicant.

(*b*) To provide the applicant with enough information so that a reasonable idea of what is involved can be obtained.

The interview should be seen as a guided conversation and exchange of views rather than as a one-sided cross-examination and should be conducted in a relaxed manner as it must be remembered that there is often tension and apprehension on the side of the applicant. Some sensible applicants carry with them a short list of questions which they will ask before leaving.

The *basic facts* required are:

(*a*) Age, date, and place of birth. Is a work permit required? Is there the question of pension and superannuation rights? Is the applicant too near retirement age?

(*b*) Previous positions and the length of time in them. Is the experience similar to the job applied for? Is it reasonable to suppose that the applicant will be permanent or does the work history indicate that the employment will only last for a few weeks?

(*c*) Qualifications. Are they of a suitable standard?

(*d*) Health and sickness record. This is of particular importance when engaging staff for kitchen, restaurant, or dining-room work; in some cases, it may be the policy of the establishment to insist on a medical examination before engagement. Varicose veins are not suitable for jobs where long standing is involved, however willing their owners may be.

(*e*) Home and holiday commitments. What arrangements are there for small children? Does the husband work shift duties so that differing working hours are required?

(*f*) Reasons for leaving the previous employment. Can one apply to them for a reference? Is there anyone else to refer to?

(*g*) And, of course, the full name and address of the applicant.

During the interview the general appearance, speech, and manner are noted to decide whether the applicant is suitable for the job concerned and if she is likely to fit in with the staff already employed.

CONDITIONS OF EMPLOYMENT

The applicant should want to know:

1. The terms of employment

These are set down in the Contracts of Employment Act 1972 and are:

(*a*) The scale or rate of remuneration, or the method of calculating remuneration. (*b*) The intervals at which remuneration is paid. (*c*) The normal hours of work, and any other terms and conditions relating to hours of work. (*d*) Holidays and holiday pay. (*e*) Terms and conditions relating to incapacity for work due to sickness or injury, and sick pay. (*f*) Pensions and pension schemes. (*g*) The amount of notice of termination to be given by the employee and the employer. (*h*) Statutory rights of the employee in relation to Trade Union membership and activity. (*i*) Procedure to be followed if there is any grievance about the employment so that the employee knows whom to see and what consequent action he can take if he is not satisfied.

The Act gives the right to minimum periods of notice to terminate employment and an employer is required to give one week's notice if there has been continuous employment for 13 weeks or more, two weeks' notice for employment of two years or more, four weeks' notice after employment of five years, six weeks after ten years, and a minimum of eight weeks' notice after he has been continuously employed for fifteen years. On the other hand, the employees are only required to give one week's notice if there has been continuous employment for 13 weeks or more, and this length of notice does not increase with length of service. These are the minimum legal periods and the individual contract made between employer and employee may stipulate a longer period. The Act also lays down that an employee normally employed for over 21 hours must be given a written statement about his terms of employment not later than 13 weeks after his employment has begun.

2. Any provision for uniform, meals on duty, or other services

Is uniform laundered free, if so, how often? Are deductions made for meals or are they considered part of the job? Are there any other special requirements?

3. Any residential requirements

What type of accommodation is offered? Is there a staff room? Are there restrictions on visitors and on the hours for coming in at night? What deductions are made? Is any reimbursement made when away on holiday?

It is always sensible to take the applicant to see the accommodation and the future place of work.

If it is not possible to tell the applicants the result of the interview at the time, they should be told when they are likely to hear. When engaged, the applicants will have to produce their National Insurance Card and the P45 Form from the previous employers for Income Tax purposes.

A record of the interview is normally kept, however roughly, for future reference. Most people consider that they are good interviewers and good judges of character, but the only sure way this can be proved is by following the applicant's subsequent career in the firm to see whether expectations have been realized; and, of course, one never knows how good the rejected applicants might have been. A successful interviewer can be judged by the rate of turnover of staff, whilst always remembering that this might equally be a sign of bad induction and management as well as of bad selection.

Other points to note when interviewing are:

(*a*) The halo effect, when a similar background, school, or acquaintance is discovered.

(*b*) The association of particular features, colouring, or dress with undesirable qualities.

(*c*) The desirability of thinking objectively about the job and the applicant so that relevant qualities only are considered.

(*d*) Not to make the job too attractive and not to sound too perfunctory.

Most interviewers are usually wary of any applicant who offers to do any job and to work any hours.

To reassure all prospective interviewers it should be mentioned that all people make some mistakes but, as with all things, one improves with experience. There is a much-quoted instance of an industrial study carried out by W. D. Scott in which 36 applicants were interviewed by 6 experienced personnel managers who ranked the applicants in order of their sales ability. The results showed wide disagreements and, in the case of 28 of the applicants, managers disagreed as to whether the individual should be placed in the upper or lower half of the grouping.

It seems agreed by Industrial Psychologists that interviewing is one of the least reliable ways of selection, apart from the judgement of social skills and acceptability, but until aptitude tests are introduced for each particular job – which is very unlikely in the near future – it is the method that we use. The results are not always too disastrous.

Great use is made when interviewing of Job Descriptions.

JOB DESCRIPTIONS

Their purpose is to clearly define the limits, responsibilities, and the authority, of a particular job, and they are required when engaging staff or when evaluating their place in the firm. As no job is constant the job description should be revised at stated intervals; in this way, it serves as a means of staff appraisal as the opportunity is then given for discussion of the job, its problems, and ways in which improvements might be made. By regular discussion at four- or six-monthly intervals, the objectives and aims of the job can be restated or reset without the feeling by staff members that the boss is 'picking on me' or has his favourites.

As an example, the following is a job description for a receptionist which is being re-drawn when the present holder retires. She was originally engaged to welcome guests and help with the correspondence but now also handles the money and banking. The questions asked are what are the right qualifications for this? Is the wage sufficient to compensate for the extra responsibility that has gradually been added to her duties? Now that she is leaving and being replaced, will it be necessary to employ someone with book-keeping knowledge in addition to the typing experience and the courteous manner which were the only requirements previously wanted. The job description is as follows:

Job title: Receptionist
Place of work: Industrial training centre
Scope and general purpose of job: Responsible for the welcoming

and allocation of accommodation for students, lecturers, and guests. The administration of accounts and the upkeep of the imprest account

Responsible to: Domestic Bursar

Description of duties: Is in charge of the Reception Office between 9.15 a.m. and 6.0 p.m. Is responsible for welcoming and receiving students, lecturers, and guests, allocating their rooms, instructing the porters in the disposal of luggage, and arranging transport as required. The upkeep of students' and visitors' books. The presentation and payment of visitors' bills. The recording of these. The imprest account and banking of cheques and money. The sorting of in-coming mail. The recording of out-going mail and the postage book. Dealing with all queries from students and guests. The office correspondence.

Occasional duties: Flower arrangements for formal dinners.

The qualifications required are: to O level standard, preferably in English, Arithmetic, and a Foreign Language. Book-keeping and Shorthand/typing.

The general requirements or personal specifications are: a good standard of health – good appearance, turnout, cleanliness, voice, and manner – social experience and poise – a good telephone approach – must be reliable and prepared to accept responsibility.

A good job specification makes the job of an interviewer very much easier as the more that is known about the job, the more likely he is to fit the right person into the right position.

Job descriptions are not meant to be rigid or to lead to such statements as 'that's not my job'. A well-written description covers the main aspects and should allow for most occurrences, particularly as it is not unknown to have as a last item: 'and such other duties as may reasonably come within the scope of the job'.

STAFF RECORDS

All departments – housekeeping and others – must keep some form of staff records. Even if all information can be obtained from Personnel or the Finance Department, these cannot always be depended on, as availability is restricted to a 5-day, 9 a.m. to 5 p.m., week, whereas the Housekeeping Department operates for most of the week's 168 hours. Moreover, a busy personnel department does not like being troubled too often by being asked to provide 'Mrs X's insurance number which she's forgotten but must have before she can claim her sick pay.'

A card index is normally used in the form shown in Figure 7.1. Holiday dates are sometimes added to individual cards; but it is simpler to record them together for all staff or by departments. The Wage Department must be notified of holiday dates in sufficient time to allow for holiday pay entitlement to be calculated. A specimen holiday roster is shown in Figure 7.2.

Staff record card (front)

Surname .... Forenames ....
Married/single .... Date of birth ....
Date of employment .... Rate of pay ....
Position ....
Qualifications ....
Experience ....
....
....
Previous employer ....
References ....
Date terminated .... Nat. insurance No. ....

(reverse)

Holiday entitlement ....

Reasons for leaving ....

Reference ....

FIGURE 7.1 Staff record card

Sickness is recorded by a duplicated copy of the sickness certificate; one copy will be sent to the Finance Department for their records and any pay adjustments, whilst the second copy is attached to the record card.

When the employee leaves, the leaving date is added to the second card with the true reason for leaving, if it is known, and also a brief reference. It is not always easy to remember Mrs. Y when asked for a reference six months after she has left, so if the main points are noted at the time of leaving there need be no delay in replying, even if you are on holiday or have yourself left.

Holiday roster – May/Sept 19..

Department / Housekeeping (W/e)	10/5	17/5	24/5	31/5	7/6	14/6	21/6	28/6	5/7	12/7	19/7	26/7	2/8	9/8	etc
Mrs Andrews				←	→										
Mrs Birch											←	→			
Miss Cook						←	→								
Miss Eden															
Mrs Fox								←	→						
Mrs Grainger		←	→												
Mrs Hood					←→										
Porters															
Mr Ingle															
Mr Jenks										←	→				
Mr Lambert							←	→							
Mr Moxham				←→									←→		

FIGURE 7.2 Holiday roster

A periodic check through staff records indicates future retirement patterns and the need for training for succession. Needless to say, all staff records are confidential and must be kept so.

REFERENCES

References must always be taken up as the employer has the responsibility of safeguarding both his own and his guest's property and he must, as far as he is able, ensure that all employees are trustworthy. Where employees habitually handle large sums of money a fidelity bond is often required; this might apply to a hotel receptionist, cashier, or storekeeper.

A reference is a confidential statement given by one employer to another stating the character of an employee. It can be either written or verbal. If local, a telephone call is quick and can be more revealing than a stereotyped letter which states that 'Mary has been employed by me for the past 20 years and has always given good service'. For more senior staff, a written reference is normally called for, often on a form which particularizes the qualities the prospective employer requires.

A testimonial is an 'open' reference which is given to an employee on leaving an employment and is usually addressed to 'Whom it may concern.' These are not very reliable as they can easily be manufactured on any imposing-looking writing-paper.

An employer is under no legal obligation to give a reference; he may not wish to do so because he does not want to lose a good employee. On the other hand, a willingness to give a reference may mean that he wants to dispense with the employee's services. If a reference is given, the employer is legally required to give a correct indication of ability and trustworthiness as far as he knows it. He must not give information which is incorrect or misleading.

LABOUR TURNOVER

The percentage rate of labour turnover is calculated by using the following formula:

$$\text{Turnover} = \frac{\text{Number of leavers in the period}}{\text{Average number employed in the same period}} \times 100\%$$

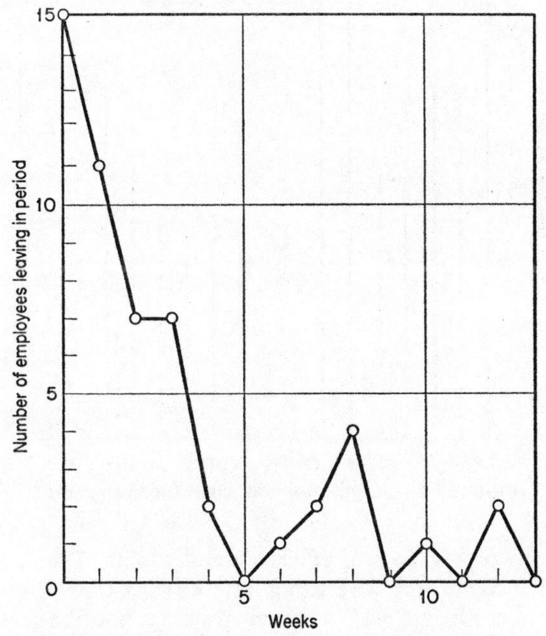

FIGURE 7.3 Labour turnover – the survival curve
In the three-month period 82 new employees have been engaged; of these, 11 left during the first week of service, 7 in the second week and the remainder as shown on the graph. By the end of the period, 37 employees had left, 45 were still employed.

This rate for labour turnover gives a useful standard of comparison for one year with another. It is calculated as the turnover of permanent staff, which are full or part-time, usually for a set period of a year, but particularly in seasonal areas, this may be for a shorter period.

Another method of measuring the success in retaining staff is the drawing of a survival curve (Figure 7.3) which shows the total number employed during the period and the numbers leaving in each week or quarter of the year.

A third method is to compile a length of service chart (Figure 7.4).

It may be difficult to arrive at the true cost of labour turnover, but some attempt should be made so as to fix standards of comparison

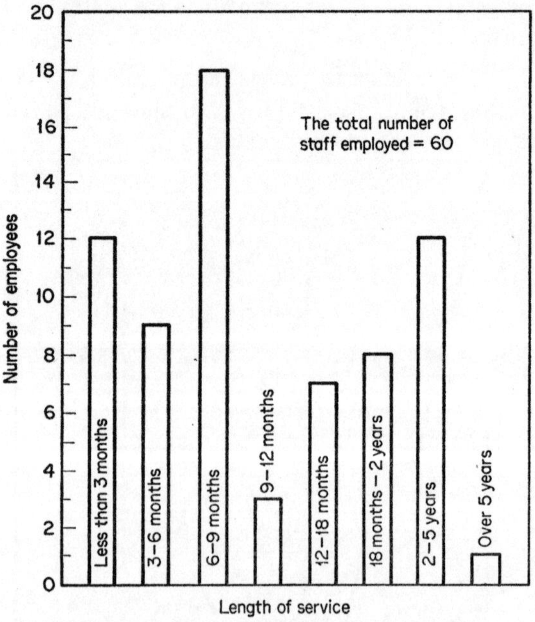

FIGURE 7.4 Length of service chart (or histogram)

and to relate cost to loss of efficiency and profits. The costs to be taken into account are: advertising, interviewing time, the administration for records and pay, uniform, training, and the loss in efficiency whilst the new employee is learning the methods and routines required.

Assessment of costs will vary from firm to firm. The calculations in Figure 7.5 give some indication of the extent of the problem.

The cost of recruiting a semi-skilled worker (female) who leaves after 12 weeks might be calculated as follows:

		£
Administration costs:		
advertising, interviewing,		
discussion, entering of	=about 2 hours	
details on record and wage cards.	+advertisement	3·00
Uniform		3·00
12 weeks wages @ £12·00		144·00
Employers contribution for Nat. Insurance		
S.E.T., etc. £1·95		23·40
Training – approx. 6 hours @ £0·50 per hour		3·00
Administration time for 12 weeks		3·00
		£179·40
Value to Firm in terms of work/output:		
1st week	£4	
2nd week	£5	
3rd week	£7	
4th week	£7	
5-12th week	£80	£103·00
Estimated loss to the Firm =		£76·40

FIGURE 7.5 The cost of labour turnover

Some movement of staff is of benefit and should be encouraged, particularly with younger skilled or semi-skilled staff as by this means they gain experience of different working methods and responsibilities and in staff handling. This encourages the interchange of ideas and methods, but the constant movement of unskilled staff who change jobs for no useful purpose is time-wasting and expensive.

Labour turnover can never be completely eliminated as there must always be a natural wastage through retirement, sickness, and the unavoidable domestic problems. The aim should be to stop unnecessary wastage, whether it is caused by bad selection, inadequate induction, or general indifference to the working conditions and needs of the staff.

TRADE UNIONS, WAGES COUNCILS, AND WAGE RATES

Since the eighteenth century organizations of both employers and employees have been developing into the present-day trade unions and professional associations. The Ministry of Employment defines a Trade Union as an organization of employees – wage-earners, manual, salaried, or professional – which includes amongst its functions that of regulating the conditions of employment of its members by collective bargaining and agreement. Since the 1870s, it has been recognized by the Government that collective bargaining is the normal means of agreeing wages and working conditions through Joint Industrial Consultation and Negotiating Committees which have equal Union and Employers representation.

Collective bargaining protects the worker by introducing a third party to negotiate on his behalf who is often in a better position to evaluate the work and conditions prevailing than the worker himself. Bargaining protects the employer from undercutting of wages by other competing firms and excludes the necessity of making individual bargains and arrangements with each employee.

Other services which are offered by the Union are legal advice, education, and training. To advance the interests of their members, Trade Unions can institute and take part in political activity but members – under the Trade Disputes Act of 1946 – are able to contract out of the payment of levies for political purposes if they so wish.

The four main types of union are craft, industrial, general, and professional.

In the Hotel and Catering Industry, Trade Union representation is fragmentary as there is, as yet, no industrial or craft union dealing specifically with catering; some sectors of the industry are more organized than others and are represented by one or other of the general or industrial unions.

In the U.K., when an industry is unorganized in this way, minimum wage rates and conditions of employment are fixed by *Wages Councils*. These are established under the Wage Council Acts of 1945, 1948, and 1959.

Employers and employees are equally represented on the Councils with a chairman appointed from one of three independent persons who are appointed to contribute an outside unbiased view to the meetings.

Minimum rates of pay and conditions of service for each grade of staff are laid down; although less may not be paid, there is nothing to stop an employer offering more pay and better conditions, and this is often done to keep rates level with those prevailing in the locality.

Factors which are considered when determining wage rates are:

(*a*) The value of the job in accordance with its importance or rank in the work structure of the organization. When an evaluation is done, various jobs fall into groups in order of degree of skill, responsibility, and importance. All jobs must be evaluated together so that a progressive salary scale can be built up and increments for length of service, qualifications, or special factors inherent in the job, can be added. Job evaluation can only be done when job descriptions are available and comparison can be made of the training requirements, knowledge, and experience needed.

(*b*) The general conditions prevailing in the industry, whether there is under- or over-employment, or if there is an increase or decrease in trade.

The age, experience, qualifications, and training of the applicant should only affect the position on the wage range for the particular job which is offered, and should be related to the pay structure of the firm. Wage rates should reflect the value placed on the job, not the value of a particular applicant.

STAFF WELFARE

It is only the very largest establishments which have a trained nurse on the staff although all hotels and residential establishments will have a doctor who is readily available. Residential staff should register with their own doctor and, when needed, consult him in the normal way.

All senior staff, Domestic Bursar's or Housekeeper's, should be capable of dealing with first aid and with sick nursing as part of their duties. Adequate first aid equipment should be provided and must be readily available. The St John Ambulance Association and the British Red Cross Society both run courses and issue certificates in First Aid and in Sick Nursing.

COMMUNICATIONS

Even in the best-run establishments, communication can lead to problems. The more dealings one has with staff, the more it is realized how clear, explicit, and definite one should be. This is not to say that staff are unintelligent, but it is incredibly easy for the simplest of actions or instructions to be misinterpreted or misquoted by a group of people. One should always adopt a friendly, courteous attitude to staff; it never helps to be abrupt or off-hand, or to shout. When giving instructions it is best to:

(*a*) Think carefully so that one defines exactly what is required.

(*b*) Use clear simple English.

(*c*) Not put too many new ideas in one sentence.

(*d*) Check that the instruction is understood by asking questions and by giving the staff the opportunity to ask questions in turn.

(*e*) Phrase the instruction in the form of a request 'Would you . . . or will you be able to . . . ?'

There is nothing wrong in repeating the instructions but there can be a lot wrong in the manner in which it is done. One should try to put oneself in the other's shoes.

If giving instructions to be passed to other members of staff, it is best to put all requirements in writing, either on the staff notice-board or in the form of a circular to those concerned.

A good Supervisor needs to be accessible so that all genuine problems can be discussed, whether these are difficulties arising from the work or from a crisis at home. It is not possible to solve every problem but, at least, understanding and sympathy can be given. The Supervisor must be careful, however, not to acquire a reputation for wanting to know all about the private concerns of the staff.

Reprimanding or correction should always be done in private with an attempt made to understand all aspects of the points at fault and to reach a solution. When misunderstandings have occurred, it is the Supervisor, Housekeeper, or Domestic Bursar who must take the initiative towards reconciliation; however much injured pride and indignation there may well be, it does not excuse a lack of leadership and control. It must also be remembered that one cannot always support a colleague against a subordinate, regardless of the truth for, if this does happen, there will be strong resentment on the side of the staff and a quick deterioration in staff relationships.

DUTY ROSTERS AND HOURS OF WORK

A considerable number of hotels and residential establishments require some categories of staff to be either on duty or on call for the full 168 hours of a week; others may only need staff for about a hundred hours each week. An average working week is from 40 to 46 hours; this is normally exclusive of meal breaks, but includes the time allowed for tea and coffee breaks. Hours are calculated by the week or over a two-weekly period, with either one and a half or two days off each week, or with the employee working a straight ten or eleven days followed by three or four consecutive days off.

Many establishments have to resort to some system of shift work or to 'split duties', and duty rosters (Figure 7.6) are drawn up to ensure that all periods are covered and to give advance notice to staff of the times when they will be required for duty.

When work schedules are prepared the following points must be considered:

(*a*) The maximum hours of work for each employee during the week.

(*b*) The maximum spreadover of hours and the hours or days when demand for staff is highest.

(*c*) The provision of meal breaks for staff and for the necessary relief staff at this time.

(*d*) An equitable arrangement for early and late duties amongst staff and a fair assignment of days off, if these do not all fall at week-ends, so that some consideration is given to personal and family circumstances.

	Mon	Tues	Wed	Thurs	Fri	Sat	Sun	Mon	Tues	Wed	Thurs	Fri	Sat	Sun
Porter 1	A	A	A	A	B	B	B	C	C	C	C	D.O.	D.O.	D.O.
Porter 2	D.O.	D.O.	D.O.	A	A	A	A	B	B	B	C	C	C	C
Porter 3	C	C	C	C	D.O.	D.O.	D.O.	A	A	A	A	B	B	B
Porter 4	B	B	B	C	C	C	C	D.O.	D.O	D.O	A	A	A	A

FIGURE 7.6 Duty roster

By this means, there will be four porters on duty each Thursday. The extra porter will be available for relief duties elsewhere or to cover special jobs in the porter's work schedule.

***Shifts*: A 6.30 a.m. – 2.30 p.m.; B 8.30 a.m. – 6.0 p.m.; C 2.0 – 10 p.m.**

(*e*) When overtime is to be worked, this must be distributed evenly between all employees.

(*f*) The possible employment of relief staff to cover days off or part-time staff for early, evening, or week-end duties to avoid split duties or overtime.

Overtime is paid for all hours worked beyond the normal working hours and for hours worked on the normal rest day. For part-time workers overtime is paid either for the hours in excess of those for which they were engaged, or for hours in excess of the normal working if the employee had been engaged for full-time work – whichever was agreed or stated in the employees' contract of employment.

Relief staff are usually employed in the larger establishments to cover periods of holiday, sickness, and the days off which are not covered in the department. A relief worker should be very adaptable and is often more highly skilled than the other workers although there is a tendency to place new employees on relief work before it is decided where the most suitable position for them will be.

When extra payments have to be made for Saturday or Sunday work, it is frequently the policy of the firm to reduce this working period to the minimum.

Hours of work and other conditions of employment for young people are controlled by the Young Persons (Employment) Act, 1938 and by the Shops Act, 1950. An example of a duty roster follows.

DUTY ROSTER FOR FOUR PORTERS EMPLOYED IN A PROVINCIAL TOWN

The hours to be covered by the day porters are from 6.30 a.m. to 10.0 p.m. Night staff are employed from 10.0 p.m. to 6.30 a.m.

The porters work an 84-hour fortnight with 3-shift working

Shifts:	*A*	*B*	*C*	
		Porter No.		
Monday	2	3	4	Days off for Porter No. 1
Tuesday	4	2	3	
Wednesday	3	4	2	
Thursday	2/4	–	1/3	
Friday	1	3	4	Days off for Porter No. 2
Saturday	4	1	3	
Sunday	3	4	1	
Monday	1/3	–	2/4	
Tuesday	2	4	1	Days off for Porter No. 3
Wednesday	1	2	4	
Thursday	4	1	2	
Friday	2	3	1	Days off for Porter No. 4
Saturday	1	2	3	
Sunday	3	1	2	

Hours for each shift: A – 6.30 a.m. to 2.30 p.m. = 8 hours
B – 8.30 a.m. to 6.0 p.m. = 9½ hours
C – 2.0 p.m. to 10.0 p.m. = 8 hours

All porters work: 4 × A shifts = 32 hours
3 × B shifts = 28½ hours
4 × C shifts = 32 hours = 92½ hours
Deduction for meals taken = 8½ hours
Total hours worked: = 84 hours

On alternate Mondays and Thursdays, one porter could be used as a relief for other duties.

An alternative method might be arranged with the porters working 4 A-shifts, 4 C-shifts and 3 B-shifts together as is shown in Figure 7.6.

TRAINING

Training has always been necessary, particularly with domestic staff who are willing to learn but who expect everything to be done in the same way that it is done at home. About 90% of the money spent on cleaning is for labour and only about 10% is spent on cleaning materials and equipment so that it seems reasonable to train staff in the most economical and best methods of working. An added incentive has been provided by the Industrial Training Act of 1964 with its three main objectives. These are:

1. To ensure an adequate supply of properly trained men and women at all levels in industry.

2. To improve the quality and efficiency of industrial training.
3. To share the cost of training more evenly between the employers in the industry.

The Hotel and Catering Industry Training Board was set up in 1966 and, by a levy/grant system, the employment of Training Advisers, and the formation of Group Training Schemes, is beginning to standardize and encourage industry-based training.

Most large firms have now finished with the old 'learning from Nellie' technique and are introducing training schemes for all grades of staff: craft, supervisory, and management. As well as basic 'on-job training', instruction is particularly necessary when new equipment is introduced, when high standards of work are required, and where there are special problems with cleaning and hygiene, such as in the Hospital Service where there is the constant risk of cross-infection. Foreign staff may need to become familiar with the English language and with the use of common terms as well as with unfamiliar types of equipment.

INDUCTION

A basic induction programme and points to be covered for all staff which might be followed in a large hotel or other residential establishment is as follows. This would be introduced as a routine procedure for all newcomers on their first day of employment.

INDUCTION

Brief introduction to the hotel, its aims, and management structure.

The general approach to the customer

Attitudes are important; always be pleasant, respectful, and polite. When meeting customers say 'Good morning Sir/Good morning Madam'. A smile works wonders. There is no such thing as *an awkward customer: but* he may be tired, worried, and irritable; if there is anything you can do to help, you may have gained a friend for the hotel, a customer who will remember and will want to return when he is next in the district.

What the newcomer must know:

1. *Conditions of Work*

(*a*) *Uniform* – what is correct, where and when it must be changed. The importance of always looking fresh, clean, and neat.

(*b*) *Pay and Hours of Work*. Clocking in/out procedure. Necessity of always clocking in/out for oneself, *never* for anyone else. Days off.

Where pay is drawn and on which day. Payments made a week in arrears.

(*c*) Arrangements for *meals and tea breaks;* cloakroom and locker facilities.

(*d*) *Absence* through sickness or other causes. Sickness certificates.

(*e*) *Holidays.* Entitlement. When and how to book holidays. Importance of giving adequate notice for particular days off.

(*f*) *Termination.* Length of notice to be given. Return of uniform.

2. *Complaints Procedure.*

Never argue but apologize to the customer for any failure in the service and try to put things right if you are able to do so. Please report all complaints to the supervisor so that action can be taken to prevent the same fault recurring.

3. *Fire Precautions.*

Routine to be followed:

(*a*) Report fire immediately to switchboard. Use of alarm system.

(*b*) Switch off all machinery. Lifts must also be switched off and not used.

(*c*) Close all doors and windows.

(*d*) When instructed, clear building with as much speed as possible but do not panic. Do not go to cloakrooms or return to the building until instructed to do so.

(*e*) Make sure you know which are the right escape routes from your department.

(*f*) Know the location of the nearest fire-fighting equipment, and how it works.

Most accidental fires are caused either by carelessness with cigarette ends and spent matches or by electrical faults; so be careful.

4. *First Aid and Accidents*

Every effort must be made to protect the customer and other members of staff from personal injury or damage to clothing.

(*a*) Know where the first aid boxes are kept and what steps to take in the event of minor cuts and burns. *Report* all accidents to the supervisor.

(*b*) Particular attention should be paid to damaged or curled carpets, splintered wood, sharp corners or protruding nails, rubbish or equipment left in corridors and entrances, and greasy floors. If these cannot be rectified by you, report the fault to the supervisor.

5. *Security*

Please safeguard the property of the customer, other members of staff, and the hotel. *Report* any suspicious or peculiar occurrence to the supervisor or management.

6. *Lost Property*

Lost property should be handed to the supervisor as soon as possible with a note stating where found and the date.

Job induction

Does the newcomer know exactly what the job requires? Is there a job description? Has the basic job training programme been arranged? When will it be carried out?

Follow up

There is no point in carrying out detailed staff induction if no effort is made to ensure that the staff remains in the job. It can take several weeks for the newcomer to settle down. After a week or ten days, find out how the newcomer likes the job: are there any problems, is encouragement needed? Are standards likely to be maintained? Will further training be necessary?

A basic job induction training programme for new **housekeeping** staff might be as follows:

1. The use and care of equipment 2. The use of cleaning materials 3. Order and method of work and time allowances. Standards 4. Simple hygiene requirements, personal and public	Training given by the Housekeeper, Asst. Housekeeper, or Floor Supervisor
5. Safety procedures. Fire, Accident. Security of rooms and guests' property	Talk given by senior member of staff responsible for safety

Talks should be short, allowing ample time for discussion and questions and opportunity for the new employee to use all equipment. Often there is not the time or the need for induction to be carried out full-time during the first days of employment but sessions can be fitted into slack afternoon periods. Induction should, however, be completed within the first two or three weeks of employment. The order of priority will vary from hotel to hotel with the exception of safety procedures which should always be dealt with on the first day.

Initial training should be reviewed at regular intervals to ensure that it is still relevant and to make sure that the methods and skills taught are being put into practice. Training is a continuous process; short refresher courses can be run throughout the year or are organized pre-season so as to maintain standards. All staff should be encouraged to increase their professional knowledge through the many short courses, evening or day, run by the local technical colleges, professional organizations, and the Hotel and Catering Industry Training Board.

Courses can be arranged on most subjects if there is sufficient demand but it does require a little exertion by both employer and employee, and off-duty times may often have to be re-arranged. This may not always be convenient but, as it should lead to a more interested and knowledgeable staff, every effort should be made. The main disadvantage is that you, yourself, may have to do some surreptitious reading to keep up with them.

To encourage good on-job and off-job training, the HCITB and the Ministry of Employment train supervisors in instruction methods so that all training is effective and purposeful.

SPECIAL PROBLEMS IN STAFFING

Although not all hotels and residential establishments require to have staff on duty for the full 168 hours of a week, most have one or other member of staff available or 'on-call' for the time when an emergency might arise. Most establishments have staff on duty to cover a sixteen- or eighteen-hour period for seven days a week. This is common to all. Below are some of the chief points to emphasize when talking about the housekeeping tasks and organizational problems in the main types of establishment. The points may, of course, be equally relevant to all sides of the industry.

HOTELS

1. Staff must be trained to be adaptable and to develop good customer relationships.
2. There must be a flexible cleaning programme to suit the guest's requirements, with the provision of early morning or over-night cleaning for public rooms.
3. Cleaning should be quiet and unobtrusive.
4. High standards have to be maintained for comfort, furnishings, and cleanliness.
5. Redecoration and periodical cleaning have to be completed in slack periods as they occur.
6. Because there is always a constant movement of guests, all staff should be aware of the policies for security, accident, and fire prevention, so that they can take the right action in an emergency.
7. A wide range of 'disposable' items, such as soap, notepaper, tissues, and shoe cleaning materials, have to be regularly replaced in each room for guests.

HOSPITALS

1. They must maintain the highest standards of hygiene and avoid the risk of cross-infection from patient to patient and ward to ward. This involves the careful training of staff and the close control of cleaning methods and equipment.
2. There is the necessity to fit the cleaning routine into the continuous and busy hospital day so that it causes the least inconvenience to patients and nursing and medical staff. In many cases cleaning must be dovetailed in with the nursing procedures; as an example, ward sweeping follows the making of the beds, should not precede it, and should not be carried out at a time when dressings are being done.
2. Staff must be flexible and able to adjust to the many interruptions that can occur. Staff/patient attitudes and the need for quietness must also be stressed.
3. Equipment and working methods should be chosen which make the least amount of noise.
4. Different working methods and standards may be needed in different departments so that the interchange of staff can be difficult.
5. Whilst having to maintain high standards, most hospitals employ a high proportion of unskilled and foreign labour who have widely differing standards; language can be a barrier and may mean the need for instruction in basic English and the translation of working methods and instruction cards into the appropriate language.
6. Difficulties can arise as the domestic staff, whilst under the control of the Domestic Superintendent, also come under the jurisdiction of Departmental and Ward Nursing Staff.

ACCOMMODATION FOR HOSPITAL STAFF

1. Hospital staff work irregular hours so that, again, cleaning routines must be flexible and off-duty periods not disturbed.
2. Where night staff are accommodated, care must be taken to prevent equipment noise, disturbance from banging doors, and corridor clatter.
3. As the accommodation is frequently used as a permanent home, staff are encouraged to have many of their own possessions; this means that staff have to take extra care with personal belongings.

HALLS OF RESIDENCE

1. These are usually occupied by students for only thirty to thirty-six weeks of the year, so staff are often paid a retainer of half

pay for the remaining weeks. This may be suitable and convenient for married women who have school children to cope with during the holidays, but it acts as a deterrent to prospective staff who need to be self-supporting.

2. To gain extra revenue, many authorities take bookings for vacational conferences and courses; because so many married women are employed staffing can become difficult. There is also the question of storage of the student's personal belongings.
3. Welfare and sickness facilities for the students frequently mean a great deal of personal involvement by the staff who must always feel some responsibility for the well-being of the students and for the good name of the establishment.
4. There can be problems concerning the security of the building and its contents in the evenings and weekends and during the vacations, particularly when few staff are on duty.
5. The use of kitchen and dining-room equipment and accommodation is often uneconomical when not fully utilized and used only for morning and evening meals and at weekends.

8 STATUTORY REQUIREMENTS

SOCIAL SECURITY LEVIES

NATIONAL INSURANCE

The National Insurance Acts of 1946 and 1966 provide for the compulsory insurance of all people who are between school-leaving and retirement ages. The population is divided into three classes. These are:

Class 1 (Employed) – for all who work for an employer
Class 2 (Self-employed) – for those working on their own account
Class 3 (Non-employed) – for all who are not employed in gainful work

Flat-rate contributions are paid by each category.
The benefits which the scheme provides are:

Class 1	*Class* 2	*Class* 3
Unemployment benefit		
Sickness benefit	Sickness benefit	
Maternity allowance	Maternity allowance	
Maternity grant	Maternity grant	Maternity grant
Widow's benefit	Widow's benefit	Widow's benefit
Guardian's allowance	Guardian's allowance	Guardian's allowance
Child's special allowance	Child's special allowance	Child's special allowance
Retirement pension	Retirement pension	Retirement pension
Death grant	Death grant	Death grant
Injury benefit		
Disablement benefit		
Industrial death benefit		

For classes 2 and 3, responsibility rests with the individual to buy the necessary stamp, affix it to the National Insurance Card and return it to the Ministry of Social Security when completed.

Class 1 contributions are made up of two parts:

(*a*) The employee's contribution which is deducted by the employer from the gross pay, and
(*b*) The employer's contribution.

A proportion of both the employer's and the employee's contributions are for National Insurance, National Health Contributions, and Industrial Injuries Insurance. The employer also contributes to the Redundancy Fund.

Insurance cards are held by the employer and it is the employer's responsibility to see that all payments are made; if this has not been done, the employee is legally entitled to claim from the employer any benefits he would have been entitled to if payments had been made.

Married women, who are covered by their husband's insurance payments, can opt to pay only the Industrial Injuries part of the contributions; this means, however, that they will not be entitled to any sickness or unemployment benefits. Most firms deduct from the gross wages the sickness benefit the employee could have received if the full insurance had been paid, whether this has been paid or not.

INDUSTRIAL INJURIES

Industrial injury benefit is paid for a period – at present 26 weeks – if the worker has been unable to work through an accident caused at work or by reason of a 'prescribed industrial disease'. Benefit is not paid for the first three days of incapacity.

The Ministry defines an 'accident' as an 'unexpected happening resulting in a personal injury whether the effects of the accident are immediate, as when someone breaks a leg in a fall, or delayed, as when a trivial wound or scratch later becomes septic'. Also 'an accident is treated as having risen out of the employment if it occurred because the injured person was doing something he was employed to do or because his employment exposed him to a special risk'.

Accident Books are kept in which all injury is recorded (*see* page 155).

Any employer who is subject to either the Factories Act or the Offices, Shops and Railway Premises Act is required by law to report to the District Inspector all accidents which have occurred on the premises resulting in an absence of more than three days (*see* Figure 8.1). When an employee claims industrial injury benefits from the Ministry of Social Security, a form will be sent by them to the employer for completion, and no payment is made until this is returned. The form asks for all relevant details of the accident such as time, place, cause, the names of any witnesses, and whether machinery was involved.

GRADUATED PENSIONS

The Graduated Pensions Act 1961 provided for all people earning between £9 and £30 each week, who were not in an industrial or

NOTICE OF ACCIDENT

Form prescribed by the Ministry of Social Security for the purpose of section 48 of the Offices, Shops and Railway Premises Act, 1963.

1. *Occupier of Premises*
   Name..
   (*a*) Address ....
   Nature of Business ....
   (*b*) Actual employer of injured person if other than above
   Name..
   Address ....
2. *Injured Person*
   (*a*) Full Name (Surname first) Mr/Mrs/Miss ....
   ....
   (*b*) Age .... Occupation.......................
   (*c*) Address ....
3. *Place Where Accident Happened*
   (*a*) Address (if different from 1(*a*) above) ....
   ....
   (*b*) Exact location (e.g. staircase to office; canteen storeroom; shop counter) ....
4. *Accident*
   (*a*) Date.................. Time ....
   (*b*) Full details of how the accident happened and what the injured person was doing. If a fall of person or materials, plant, etc., state height of fall ....
   (*c*) If due to machinery, state:
   (i) Name and type of machine? ....
   (ii) What part of the machine caused the accident? ....
   ....
   (iii) Was the machine in motion by mechanical power at the time? ....
5. *Injuries and Disablement*
   (*a*) Whether fatal or non-fatal ....
   (*b*) Nature and extent of injury (e.g. fracture of leg, laceration of arm, scalded foot, scratch on hand followed by sepsis)
   ....

Signature of Occupier or Agent ....
Date ....

Particulars additional to those prescribed:
Names and Addresses of Witnesses of Accident ....
Address to which any inquiries about this accident should be sent.
....

FIGURE 8.1

superannuation pension scheme, to receive a 'graduated' pension depending on earnings so that the final retirement pension should be more related to the employee's previous income. Arrangements were made for those firms which operated industrial pension schemes to 'opt-out' or '*contract-out*' from the State scheme. Since 1966, there has been a liability for graduated contributions from employees who have contracted out but at a lower rate than for employees who are not contracted out. This is to cover extra payments made, Earnings-related Benefits, for sickness and unemployment. Payments for graduated pension are paid and recorded with Income Tax payments on the P.A.Y.E. cards.

Flat-rate National Insurance contributions for Class 1 (Employed persons) are therefore at different rates: one for those who are *not* contracted out, and the other, at a higher rate, for those who are contracted out.

In addition to the graduated pension scaled rates, benefit for sickness and unemployment is paid at higher rates for incomes between £9 and £30, calculated on earnings.

REDUNDANCY PAYMENTS

The Redundancy Payments Act was introduced in 1965 to enable employers to make lump sum compensation payments to employees dismissed because of redundancy. This applies to all who have been employed continuously for more than 2 years and who normally work more than 21 hours each week. The Act was introduced to help safeguard the employee's interests and also to encourage employers to become more efficient by stream-lining their operations and increasing productivity. Redundancy occurs when the employer 'needs fewer employees to do work of a particular kind'.

INCOME TAX

Income tax is paid by all people in employment, by means of Pay-As-You-Earn payments wherein tax is deducted from the gross pay as it is earned.

Each employee has a code number which is calculated by the Tax Office after taking into account all income and claims for allowances which have been submitted by the employee on his tax return. Code numbers are issued so that the correct amount of tax is deducted. Any query should be referred to the Tax Office concerned.

Weekly or Monthly Tax Tables show the cumulative tax due in respect of the total wages earned within the financial year – which runs from April 6th to April 5th – and the amount to deduct from the current wages.

Tax Deduction Cards are kept for each employee which, when completed for the tax year, are returned to the Collector of Taxes to whom the amount of money collected in tax is sent monthly. The onus for collection is on the employer.

When an employee changes employment, a form P45 is made out in triplicate; this shows the gross wages for the period, the code number and the tax which has been deducted. Of the three copies, Part 1 is sent to the local Income Tax Office and Parts 2 and 3 are handed to the employee to be given to the new employer. Part 2 is kept by the new employer and from it the new tax deduction cards are prepared. Part 3 is sent, when completed, to the new Income Tax Office. Code numbers can only be altered on notification from the Tax Office.

If the P45 is not available or the employee has not a code number for any reason, tax is deducted at the rate shown in the Emergency Card Tables.

At the end of each tax year, the employer issues to each employee the form, P60, which shows the gross pay for the year and the amount deducted in income tax. It is necessary for the employee to keep this as it is also used by the Ministry of Social Security when claims are made for 'earnings-related' sickness and unemployment supplements.

DEDUCTIONS FROM PAY

When applicable, deductions for National Insurance, Graduated Pensions, and Income Tax are the only statutory deductions which can be made from an employee's pay. Other deductions can only be made with the signed consent of the employee; this applies to such things as the refunding of an advance of pay, Saving Schemes, a Holiday Fund, and Trade Union subscriptions. Payments of wages and deduction of pay for manual workers are governed by the *Truck Acts* 1831–1940 and the Payment of Wages (by Cheque) Act 1960.

PAYMENT OF WAGES BY CHEQUE

From 1963, manual workers may be paid by cheque provided there has been a written request for this by the employee concerned. No manual employee can be required, as a condition of employment, to receive wages by cheque; he must be entirely free to decide. The employer has the option after receiving the request of either paying by cheque or giving *written* notice to the employee that he will do so in the future or that he does not agree to do so.

It is obviously much simpler to pay all wages by cheque but unless there is a unanimous decision on the part of the employees no employer will want to have to pay only part of the wage-sheet in this way.

OTHER LEGISLATION

Other legislation which the housekeeper should be familiar with is as follows but where there is any doubt as to interpretation or relevance, the Act itself should be studied or else a legal adviser consulted.

1. *EMPLOYER'S LIABILITY (DEFECTIVE EQUIPMENT) ACT 1969*

This applies when there has been a 'defect in the equipment' which has been supplied by the employer for the purpose of the employer's business and an employee has, in the course of his employment, been injured as a result of the defect. If the equipment has been defective when supplied by the manufacturer, the supplier can be brought into the case as a third party, provided he is not insolvent, out of business, or abroad; otherwise, it is the employer who is liable to be sued for damages by the employee; which gives grounds for a full comprehensive insurance policy on behalf of the employer.

2. *HOTEL PROPRIETORS' ACT 1956*

Innkeepers and hotel-keepers have always had a strict liability to safeguard the property of their guests. An inn is defined as any establishment which offers food, drink, or sleeping accommodation, without special contract, to any traveller who is in a fit state to be received and is able to make payment for the services provided. 'Private' hotels which reserve the right to choose their guests and do not hold themselves out to receive all travellers are not under this strict liability to safeguard guests' properties, although under common law they must take all reasonable care. Under the 1956 Act, all 'common inns' are now referred to as hotels.

Liability depends on whether or not a notice is displayed. This limits the hotel-keeper's responsibility to £50 for any one article with a maximum of a total of £100 for any one guest, with the exception of any property which has been left with the hotel-keeper for deposit or safe custody. Liability only extends to the property of guests who have engaged sleeping accommodation and does not cover motor-cars or other vehicles and their contents or horses or other live animals. A specimen notice dealing with loss or damage to guests' property is shown in Figure 8.2.

Unless this notice is displayed prominently near the reception desk or office where the contract for accommodation is made, the hotel-keeper is liable for the full amount of the loss unless the guest himself can be proved to have been negligent.

All hotel-keepers and those in charge of other establishments have the common law duty to exercise reasonable care of all property.

NOTICE

LOSS OF OR DAMAGE TO GUESTS' PROPERTY

Under the Hotel Proprietors' Act, 1956, a hotel proprietor may in certain circumstances be liable to make good any loss of or damage to a guest's property even though it was not due to any fault of the proprietor or staff of the hotel. This liability, however:

(*a*) extends only to the property of guests who have engaged sleeping accommodation at the hotel;

(*b*) is limited to £50 for any one article and a total of £100 in the case of any one guest except in the case of property which has been deposited, or offered for deposit, for safe custody;

(*c*) does not cover motor cars or other vehicles of any kind or any property left in them, or horses or other live animals.

This notice does not constitute an admission either that the Act applies to this hotel or that liability thereunder attaches to the proprietor of this hotel in any particular case.

FIGURE 8.2

3. *RACE RELATIONS ACT 1965*

Race Relations Act 1965 makes it an offence to discriminate on grounds of race or colour. It precludes advertisements for staff from particular countries or the rejection of a guest on the sole grounds of colour.

4. *TRADE DESCRIPTIONS ACT 1968*

Protects the consumer against false statements or descriptions, either written or verbal, concerning both goods and services and their prices. This means, for example, that a room advertised as giving good views of the sea, must do so; there was a recent case of an hotel, advertised as modern, being summoned under the Act because, although all amenities were modern and up-to-date, the hotel itself was built some 80 years earlier and could not, as such, be so described.

5. *ALIENS ORDER 1953*

Lays down that all people over 16 years of age who stay for more than one night in any premises in which sleeping accommodation is offered for reward *must* inform the management or the proprietor of their full name and nationality. If the guest is an alien, the record must also show the passport particulars or those of any document which establishes his identity, the date of his departure, and his destination.

Records must be kept for at least one year from the date of arrival of all guests, and they are open to inspection by the police or other authorized person.

6. *OCCUPIERS' LIABILITY ACT 1957*

By this the occupier of premises has a duty to see that they are reasonably safe and that all reasonable care has been taken to protect from harm any visitor to the premises. The occupier will only be liable if it can be proved that he knew of, or should have known of, any danger. This obligation is not personal to the occupier but is considered to be delegated to his employees so that he is not able to plead that he has had no knowledge of the danger if it has been common knowledge amongst his staff.

7. *MISUSE OF DRUGS ACT 1970*

In this, a landlord or person 'concerned in the management of premises' has certain responsibilities if it is found that residents or guests are using drugs, as it is an offence to permit such use. He will not, however, be generally considered liable if drugs are found on those premises unless he 'knowingly' allowed the premises to be used for the smoking or taking of drugs.

This means that if management become aware that drugs are being used or kept in the establishment, action must be taken to stop their use and to have them removed from the premises.

Under the Act, the police have the power to search any premises, person, or vehicle when they have 'reasonable grounds' to suspect that a person is in possession of controlled drugs. They also have the power to arrest without a warrant.

9 WORK STUDY

AIM AND APPLICATION

WORK study is one of the tools of management. The aim is to provide information on the working situation so that the most effective use of equipment, space, and human effort may be made. Work study can be divided into two main branches: Method Study and Work Measurement.

Traditionally 'Motion Economy', or the application of Methods Analysis at the workplace, forms a sub-division of Method Study. Closely allied and important for the application of these work study techniques is the study of human work in relation to the working environment or, as it is more concisely referred to, the Study of Ergonomics.

One of the greatest difficulties in introducing work study techniques is that these practices are often fiercely resented by the employees who fear that they may lead either to unemployment or to transfer to another job, or that they will be required to do considerably more work in the same amount of time at the same rate of pay. There is often the feeling that the work study man cannot possibly appreciate all the difficulties and problems involved and has not fully understood all the intricacies of the job. Most of us have a strong built-in resistance to change, dislike being watched closely whilst working, and do not like to think that the way in which a job has been carried out for many years has only been partially effective or that the job itself may even have been unnecessary.

Success depends on the co-operation of all staff, so it is essential that there should be detailed advance planning, good staff communications, and consultation with any Trade Unions whose members are concerned so that everyone knows the 'why, how, when, and where'.

In principle, Work Study is a service and as such should be considered a staff appointment and not carried out as part of the duties of line management. However, where the service is not readily available (or employment of a consultant too expensive) the line manager should consider changes in work methods and procedures to save effort. An intelligent person can do quite a lot; after all, it is no more than the application – albeit organized – of common sense. *Work Measurement*, however, remains *at all times* the preserve of the fully trained person, except for Work or Activity Sampling which can be carried out by an untrained person.

The main requisites for a successful work study man are:

(*a*) Tact.

(*b*) The ability never to express an opinion which might be construed as detrimental to employees or management.

(*c*) Never to usurp the position of supervisor or management.

(*d*) To be able to discuss and ask advice with the employees and management freely.

To ensure success all working routines should be carried out in a normal way.

METHOD STUDY

This is a critical examination and recording of the existing ways of working so that a better more efficient method can be found.

It has been said that successful work study can only be accomplished in an atmosphere where there is a healthy disrespect for present ways.

As method study is expensive, it is uneconomical to apply it to a job which involves little time or labour. It is carried out when the job involves some or all of the following:

(*a*) Repetitive work and a lot of labour
(*b*) Excessive movement of stores, equipment, or labour
(*c*) Considerable wastage
(*d*) Very tiring work
(*e*) Bottlenecks

Facts are recorded by using either process charts and symbols, flow or string diagrams to show movement, or film or models to show the exact workings or lay-out of the area. The conventional symbols are:

○ = An operation when an object is intentionally changed in any chemical or physical form or is assembled or prepared in some way.

□ = Inspection, when it is examined or checked for quality, quantity, etc.

⇨ = Movement or transport.

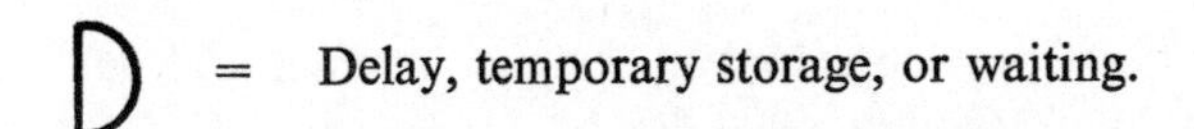

D = Delay, temporary storage, or waiting.

▽ = Storage, kept and protected against unauthorized removal.

◻ = Combined activities: operation and inspection.

Method study has to decide *what* should be done – *who* should do it – *when* it should be done – *where* and *how*? Is the job really necessary? The basic procedure is to:

SELECT
a job or procedure which can be examined with economic advantage

RECORD
by *direct* observation all the relevant facts of the present working methods – *using*

Charts
String diagrams
Films or models

EXAMINE
the facts critically and challenge all that is done – *considering*

Why? Is the job really necessary?
How? Is there a better way?
When? Is the time right?
Who? Could someone else do it better?
Where? Is it in the right place?
What? What operation has to take place?

DEVELOP
the most practical and economical method taking into account all the relevant circumstances – *considering*

New equipment
Different materials
Lay-out – stores and kitchen, cupboards, etc.
Working conditions, excessive movement

MEASURE

the quantity of work involved in the new method and agree a standard time for the job – *allowing for*

Normal speed of working – an agreed variable
Rest periods, preparation, and maintenance times

DEFINE

the new procedure so that it can always be identified – *using*

Training sessions
Job instruction cards
Work manuals

INSTALL

the new method as an agreed standard practice – *but consider*

People – that everyone concerned knows the 'how and why'
Introduce gradually; give time for adjustment, suggestions, and questioning

MAINTAIN

the new methods by regular routine checks

The study must not only take into account the actual working method but must also consider:

(a) Space saving

Rearrangement of fittings, more cupboards and benches, general lay-out and tidiness.

(b) General working conditions

Temperature, lighting, suitable equipment, sanitation and cleanliness. Welfare.

(c) Space for easy working without undue fatigue

No prolonged standing. Seats where practicable. Height of benches.

(d) Accident prevention

Do the methods conform to safety regulations? Has unnecessary movement of stores been eliminated? Level of supervision and maintenance?

MOTION ECONOMY

Frank Gilbreth, the founder of Motion Study, first used and developed principles relating to the movements of the worker at his place of work; these principles are the basis for the development of improved working methods and should be borne in mind by anyone concerned with people at work. The principles can be grouped under these headings:

1. The use of the human body so that, as far as possible, easy rhythmic movements are developed with hand and arm actions co-ordinated.
2. Arrangement of the work place with equipment, materials, and tools pre-positioned to allow the best sequence of movement so that unnecessary time and movement are not wasted in reaching or searching for requirements. Body fatigue and eye strain are prevented by the correct height of work-tables and seats, adequate lighting, and the use of colour contrasts for quick identification.
3. Design of tools and equipment so that they can be 'used with least change of body position and the greatest "mechanical advantage"'.

Amongst other points which are considered are the provision of new materials and the disposal of finished work.

With method study, as with work measurement, a decision has to be taken as to the number of observations and recordings to be made. This will depend on the complexity, length, and importance of the operation.

WORK MEASUREMENT

This provides the basic information for setting standards for labour costs and gives the means of controlling these costs by measuring the time taken to carry out an operation under normal circumstances by an average worker. To obtain a fair result, very fast or very slow workers must be avoided. The job is broken down into 'stages' or 'elements' each of which is recorded by means of a stop-watch.

The basic procedure is to:

1. Obtain and record *all* relevant information about the job, worker, and any conditions likely to affect the working situation. A check must be made to ensure that the correct method, tools, and equipment are being used.
2. Record a complete description of the job, breaking it into its 'elements'.
3. Record the time taken for each element.
4. At the same time, try to assess the speed of the worker comparing with a pre-determined 'normal' speed or pace.
5. Convert these observed times to 'normal' times.
6. Decide on the 'time allowance' to be made over and above the 'normal' time for the job.
7. By this means, determine the 'allowed' time for the job.

The I.L.O. Manual '*Introduction to Work Study*' has defined the average speed of movement of an 'average' worker as being either:

1. The normal walking pace at 3 mph over level ground and in an even temperature of about 15·6°C (60°F) which is 15 m (50 ft) in 0·189 min, or
2. The speed at which a pack of 52 cards is dealt into four equal piles in ½ a minute.

The *normal speed* should be one which can be easily maintained day after day without undue exertion and depends on the effort demanded by the job. It is affected by the care that is needed and also by the operator's mental attitude both to the job itself and to the supervision and organization of the work. A performance of 100 is considered 'normal'. Therefore a slower worker might be judged as 90 or 85 whilst a faster worker might be 110 or even 120.

	Observed Time		*Rating*		*Constant or Normal Time*
Average worker	0·48	×	100	=	0·48 min
Faster worker	0·40	×	120	=	0·48 min
Slower worker	0·60 min	×	80	=	0·48 min

$$\frac{\text{Observed time (0·40 min)} \times \text{Rating (120)}}{\text{Normal rating (100)}} = \text{Constant or normal time (0·48 min)}$$

This normalized time is, in the opinion of the observer, the time needed for each part of the job undertaken by a worker who is trained and experienced in his work. To it must be added other *Time Factors* or *Time Allowances* which will be needed for all or some of the following reasons:

1. The need to recover from the effort expended – a fatigue allowance.
2. The need for time for personal use – washing etc.
3. An agreed variable allowance to recompense for mental strain and monotony, excess noise, heat, bad lighting, abnormal working conditions, etc.
4. The time required for preparing and maintaining equipment, fetching materials, or just waiting for other work to be finished; this must be judged for each individual job.

The average worker needs adequate rest during the day if he is to keep up the required working pace throughout the year. It is usually considered that for those people engaged in comparatively light work an allowance of 12% is adequate; and for those who do heavier work, up to 20%; that is 1 or 1½ hours out of an 8-hour day. This does, of course, include the allowance for personal use. These agreed time allowances are added to the observed times to give the total time allowed for a particular job.

There has been a lot of research into how these rest pauses should be distributed throughout the day resulting in evidence that indicates that short frequent breaks seem to delay the onset of fatigue better than longer pauses between extended periods of work.

Normally, 10 to 15 minutes is allowed for mid-morning and mid-afternoon breaks when coffee, tea, or a snack are available, and the remainder of the time is taken at the discretion of the worker.

There is also evidence accumulating that maintains that a short session of P.T. or a brisk walk around the outside of the building is of more benefit than the more usual tea-break; but, although the scientists are prepared to back this belief, it is very doubtful if the average domestic worker would even consider the proposition, let alone practise it for a few days.

There are other methods of Work Measurement than the stop-watch method described above. Of these *Work or Activity Sampling* can be very useful to the institutional manager.

If management is to be efficient it is essential that it knows what is going on at the work point, to know how long a machine or an employee is standing idle or, more often, to find what percentage of an employee's time is spent doing a particular job. Employing observers is expensive and time-consuming.

WORK OR ACTIVITY SAMPLING

Work or Activity Sampling is a technique which is quickly learned and easy to apply, and one in which instantaneous observations are made and recorded over a period of time either for an operative or for a piece of machinery. The number of observations of activities such as vacuuming, dusting, polishing, and time spent waiting or idle time, indicates the percentage of time which that activity takes.

If thirty random observations are taken during an 8-hour day and in twelve of the observations the employee was vacuuming then the percentage of time for this job will be $= \frac{12}{30} \times 100 = 40\%$.

The larger the number of observations taken the more accurate the result; but for most purposes an accuracy of 95% is all that is needed; this is known as a 95% confidence limit.

The number of observations required can be estimated from the following formula:

$$N = \frac{4P\,(100-P)}{L^2}$$

where N = number of observations required
P = probable percentage of working time taken by activity being investigated
L = percentage limits of accuracy.

For example, supposing $P = 40\%$ and the accuracy required is $L = \pm 3\%$ then

$$N = \frac{4 \times 40\,(100-40)}{3 \times 3} = 1066,$$

so that if 1066 observations have been made and the result is $P = 39\%$ then it can be said that it is 95% certain the answer is between $39 - 3 = 36\%$ and $39 + 3 = 42\%$.

A nomogram may be used in which the percentage and number of observations required can be read from a chart.

Observations must be taken at random, and be distributed unevenly over all hours of the day or week, so as to avoid any cyclical activities.

The information provided by work sampling is:

1. The percentage of the working day during which employees or machines are productive and the time spent on each activity.
2. The percentage of time wasted by delay and the reason for such delay.
3. The relative activity of different employees and machines.

ERGONOMICS

This is the study of fitting the environment and tools and equipment to man and so avoiding all unnecessary strain and tension so that the best use can be made of human resources.

Consideration is given to:

1. *ENVIRONMENT*

The effect of excessive noise, the effect of heat and cold, ventilation, bad lighting conditions, and excessive glare.

2. *SPACE LAYOUT*

Space Layout which considers the effect of the size of the body in relation to the working space and heights of tables and benches. For ease and good posture, a different height of table is needed for rolling out pastry than is needed for chopping-up vegetables. In one establishment, a badly positioned sink which caused considerable stooping has been the main cause of a high turnover rate amongst kitchen porters. *See* Figure 9.1.

3. *THE DESIGN OF TOOLS AND EQUIPMENT*

Machines that are well balanced and are not inclined to run away with the lighter-weight domestic workers; machines which are safe to operate with no undue corners or switches in inaccessable positions.

Ergonomics is a recognized science, very much concerned with the

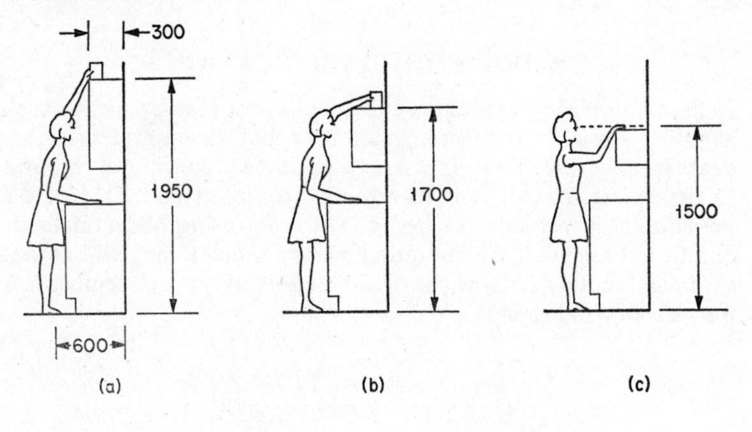

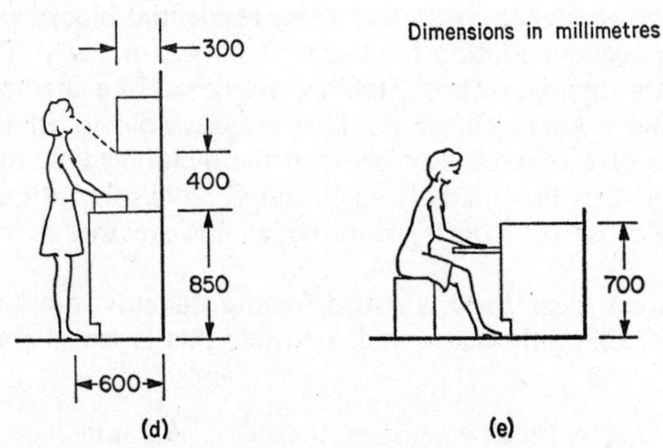

FIGURE 9.1 Working heights (assuming height of woman to be 1 630 mm)

(*a*) Maximum vertical reach over work top.
(*b*) Comfortable vertical reach over work top.
(*c*) Shelf at eye level.
(*d*) Comfortable height of work top for standing position and clearances for cupboards over. The larger dimension allows for the use of a mixer or similar appliance.
(*e*) Comfortable height of work top for seated position. Seat height 500 mm.

(From *Space in the Home* (metric edition), Design Bulletin No. 6, Ministry of Housing and Local Government, reprinted 1970.)

niceties of the ability to read different lettering on dials etc., to prevent misreading; with the effects of tedium on close concentration; with the problems of automated versus human controls, particularly at their 'interface', i.e. the point where human effort connects with machine effort. Simple *applications* of Ergonomics may be very much a matter of 'Motion Economy'; thus at the simple level no very clear distinction can be drawn between 'Ergonomics' and 'Motion Economy'.

WORK STUDY APPLICATION

In the studies below the full machinery of Work Measurement – such as rating – has not been brought to bear, but the examples demonstrate how worthwhile savings can be achieved even without recourse to the full technique. It is in any case reckoned that 80% of the possible improvements realized comes from using better methods and that the expensive elaboration of stop-watch timing will at best save only another 20% of the original effort. Always remember *it is the method that counts*!

LINEN DISTRIBUTION IN A HALL OF RESIDENCE

The hall of residence consists of three residential blocks with study-bedroom accommodation for twenty students on each of the three floors; lifts are used by both staff and residents. The linen-room is on the ground floor in Block A. This is accessible to all the blocks by means of a covered way between the buildings but, in practice, it is found that the maids from B and C blocks find it quicker and easier to cross the courtyard provided the weather is sufficiently fine.

Each week clean linen is issued for the students who are allowed one sheet, one pillowcase, and a towel; this is timed for issue as follows:

A Block .... 9.0 a.m.
B Block .... 9.20 a.m.
C Block .... 9.40 a.m.

This time, when the maids bring in the dirty linen and exchange it for clean, tends to be one of confusion as they treat it as a social occasion and are liable to linger causing congestion and delay.

Work study observations were made over a 5-week period and approximate times were taken as it was found that considerable delays were caused when the lifts were being used for other purposes.

The results have been charted and are shown in Figures 9.2, 9.3, and 9.4, which show the present and the proposed new methods. The team also considered the actual process of issue in the linen-room and this is shown in Figures 9.5 and 9.6. Figure 9.7 shows the proposed new method of issue.

Before the work study team put forward their proposals, they discussed the present system of working with the housekeeper and her staff. Amongst the suggestions which were made were the following:

1. To bring the counter back nearer to the issue shelves. This would

mean less movement for the linen-keeper and make more space for the maids with their trolleys.

2. The linen-keeper to prepare nine piles of clean laundry on the previous day and place them on the counter so that they are ready for issue.
3. To make sure that the maids arrived at set intervals so as to avoid congestion. Intervals of ten minutes were suggested as these would

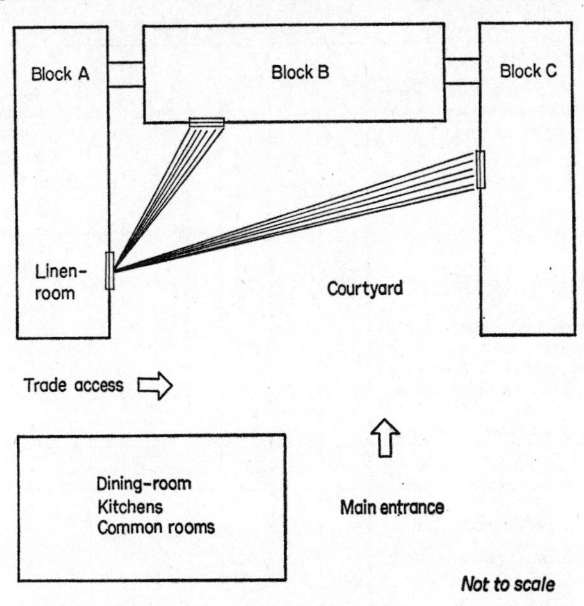

FIGURE 9.2 Distribution of linen in a hall of residence – present method

give the linen-keeper time to count and pack the dirty linen as it was received. This was strongly opposed by the housekeeper who did not consider that the linen-keeper should handle dirty and clean linen alternately; moreover it would hold up work on the blocks as the last maid would be collecting her linen at about 10.30 a.m.

4. To have the dirty linen packed and sent to the laundry from each block and the clean laundry returned to the blocks. This would have meant that the linen-keeper would have to visit each block at least twice, and extra storage facilities would have to be found. Repairs would still have to be taken to the linen-room.
5. To issue clean linen once a fortnight. The housekeeper felt that in this case the maids would have to make two journeys as they, and the small trolleys they used, would not be able to cope with the extra bulk and weight. She also raised a second point that, as the

Date *2/197-*	**SUMMARY**				
Location *Hall of Residence*	Activity		Present	Proposed	Saving
Job *Linen Distribution*	Operation	○	*9*	*11*	*−2*
Operator *① Linen-keeper and 9 maids*	Transport	⇨	*18*	*14*	*4*
	Delay	D	*9*	*–*	*9*
② Linen-keeper and Porter	Inspection	□			
	Storage	▽			
	Distance, m				
	Time, min		*235*	*110*	*125*
	Cost, £				

Description	Qty	Dist. (m)	Time (min)	○	⇨	D	□	▽	Remarks
Present Method ①									
Each maid (9) takes dirty linen to linen-room 20 sheets, towels and pillowcases					•				*Awkward load balanced on cleaners trolley weight approx 25 kg*
A1	*20*		*2*		•				
A2	*20*		*4*		•				
A3	*20*		*5*		•				
B1	*20*		*5*		•				*Delays caused by lifts in use*
B2	*20*		*10*		•				
B3	*20*		*10*		•				
C1	*20*		*8*		•				*Maids wait for each other before going to linen-room*
C2	*20*		*12*		•				
C3	*20*		*13*		•				
The average time waiting in linen-room = 8 min each			*72*			*9*			
The average time for linen-exchange = 1·45 min + 1·30 min required to load trolley.			*25*	*9*					
Return of each maid to own floor			*69*		*9*				
Total			*235*	*9*	*18*	*9*			
Proposed Method ②									
Porter collects trolley and clean linen and takes it to A1 (on ground floor)	*60*		*2*		•				
(a) Exchanges linen	*20*		*3*	•					
(b) Lift to maid 2			*5*		•				
(c) Exchanges linen	*20*		*3*	•					
(d) Lift to maid 3			*5*		•				
(e) Exchanges linen	*20*		*3*	•					
Returns to linen-room			*5*		•				
Exchanges linen	*60*		*6*	•					
To block B, and maid 1			*5*		•				
Porter repeats (a) to (e) as above	*60*		*19*	*3*	*3*				
Returns to linen-room			*10*		•				
Exchanges linen	*60*		*6*	•					
To block C and maid 1			*8*		•				
Porter repeats (a) to (e) as above	*60*		*19*	*3*	*3*				
Returns to linen-room			*13*		•				
Total			*110*	*11*	*14*				

Present Method	*Time required:*	*for collection of linen by maids*	=	*235 min*
		for issue by linen-keeper	=	*13 min*
		Total time required		*248 min*
Proposed Method	*Time required:*	*for distribution by porter*	=	*110 min*
		for issue by linen-keeper	=	*10·35 min*
		Total time required		*121 min*

FIGURE 9.3 Flow process chart – distribution of linen in a hall of residence

students made their own beds, clean sheet day was the one time in the week that she was sure that the beds were really aired and properly made.

6. To dispense with sheets, pillowcases, and blankets, and to use sleeping bags instead. This, the housekeeper considered, would be a retrograde step; although sleeping bags might be quite suitable for a settee-type of bed, the divans which were in use would be quite unsuitable.

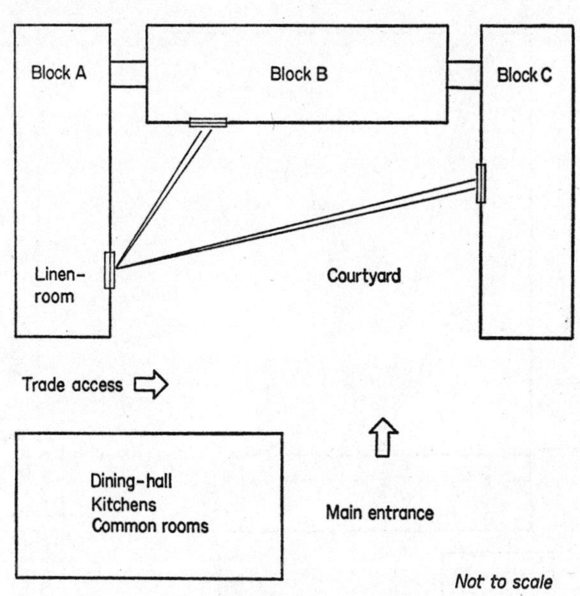

FIGURE 9.4 Distribution of linen in a hall of residence – proposed method

It was finally decided to adopt the team's proposals that a porter should be used who would distribute the sixty sets of clean linen to each block and, if all were not required, he would return the surplus articles. This would reduce the time at present required by about two hours and give the linen-keeper the opportunity to work in a more peaceful atmosphere.

The main requirements would be the services of a porter between 9.0 a.m. and 11.0 a.m. and a larger trolley. The trolley would have to carry sheets which would need 305 × 457 × 762 mm (12 × 18 × 30 in.) in space, pillowcases needing 203 × 508 × 305 mm (8 × 20 × 12 in.), and towels needing 305 × 610 × 762 mm (12 × 24 × 30 in.). There would also have to be room for sundry items such as lavatory towels

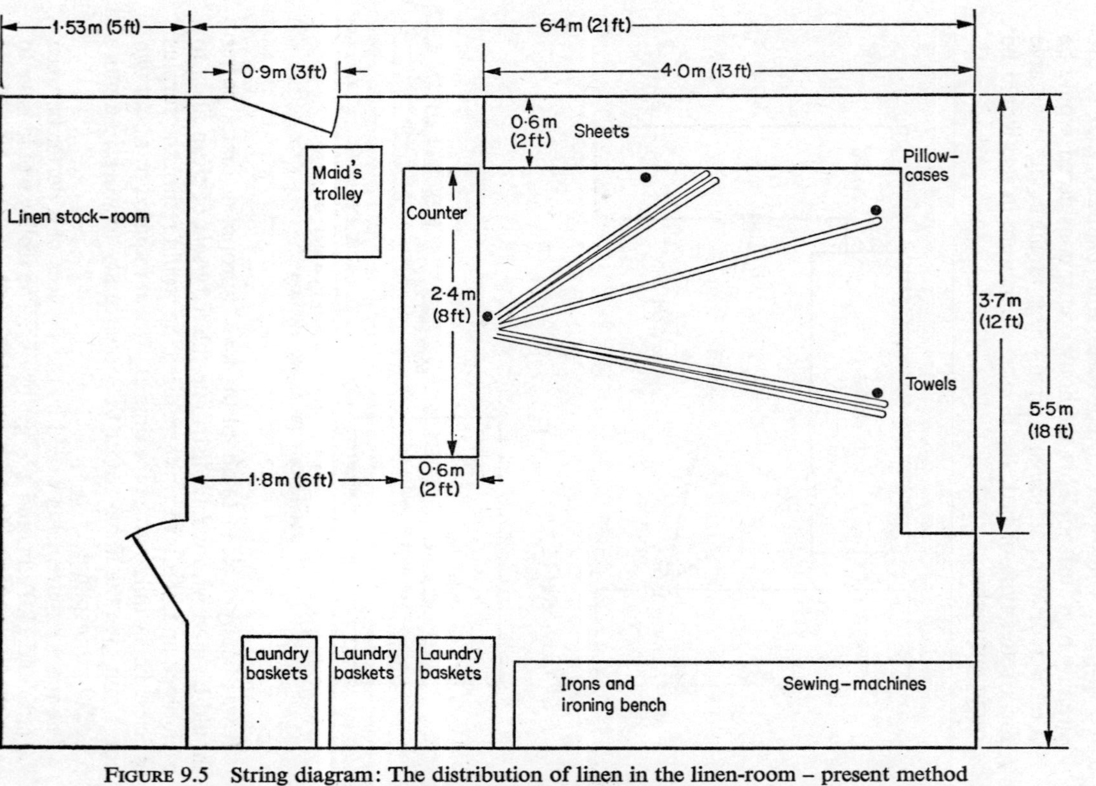

FIGURE 9.5 String diagram: The distribution of linen in the linen-room – present method
To issue 20 sheets, towels and pillowcases

Date 2/197-	SUMMARY				
Location Linen-room	Activity		Present	Proposed	Saving
Job Issue of clean linen	Operation	○	90	9	81
Operator	Transport	⇨	81	12	69
Present method—Linen-keeper	Delay	D			
	Inspection	□			
Proposed method–Linen-keeper and Porter	Storage	▽			
	Distance, m		259·2	27·9	231·3
	Time, min		13	10·35	2·5
	Cost, £		Trolley for Porter xxx		

Description	Qty	Dist. (m)	Time (min)	○	⇨	D	□	▽	Remarks
Present Method									
Linen-keeper checks issue list at counter									
(a) Goes to shelf for sheets		2·1			•				*Could maid's trolley be wheeled behind counter ?*
(b) Counts 10 sheets	10			•					
(c) Returns with them to counter		2·1			•				
(d) Places sheets on counter				•					
She repeats (a) to (d)	10	4·2		••	••				*Could the counter be nearer the shelves ?*
To shelf for pillowcases		3·4			•				
Counts 20 pillowcases	20			•					
Returns to counter		3·4	1·45						*Could the linen-keeper carry 15 or more sheets at a time ?*
Places pillowcases on counter				•					
(a) To shelf for towels		3·4			•				
(b) Counts 10 towels	10			•					
(c) Returns to counter		3·4			•				
(d) Places towels on counter				•					
Repeats (a) to (d)	10	6·8		••	••				
Totals		28·8		10	9				
The above is repeated 9 times for the issue of clean linen to each maid—not including rest periods		259·2	13	90	81				
Proposed Method									
Porter and trolley accompanied by linen-keeper to shelf for sheets	–	3·0			•				
Count 60 sheets into trolley	60	–		•					
To pillowcases		1·8			•				
Count 60 pillowcases	60	–	3·45	•					
To towels		1·8			•				
Count 60 towels	60	–		•					
Return to counter		2·7			•				
Totals		9·3	3·45	3	4				
The above is repeated 3 times for the issue of clean linen to each block		27·9	10·35	9	12				

FIGURE 9.6 Flow process chart – the distribution of linen in the linen-room

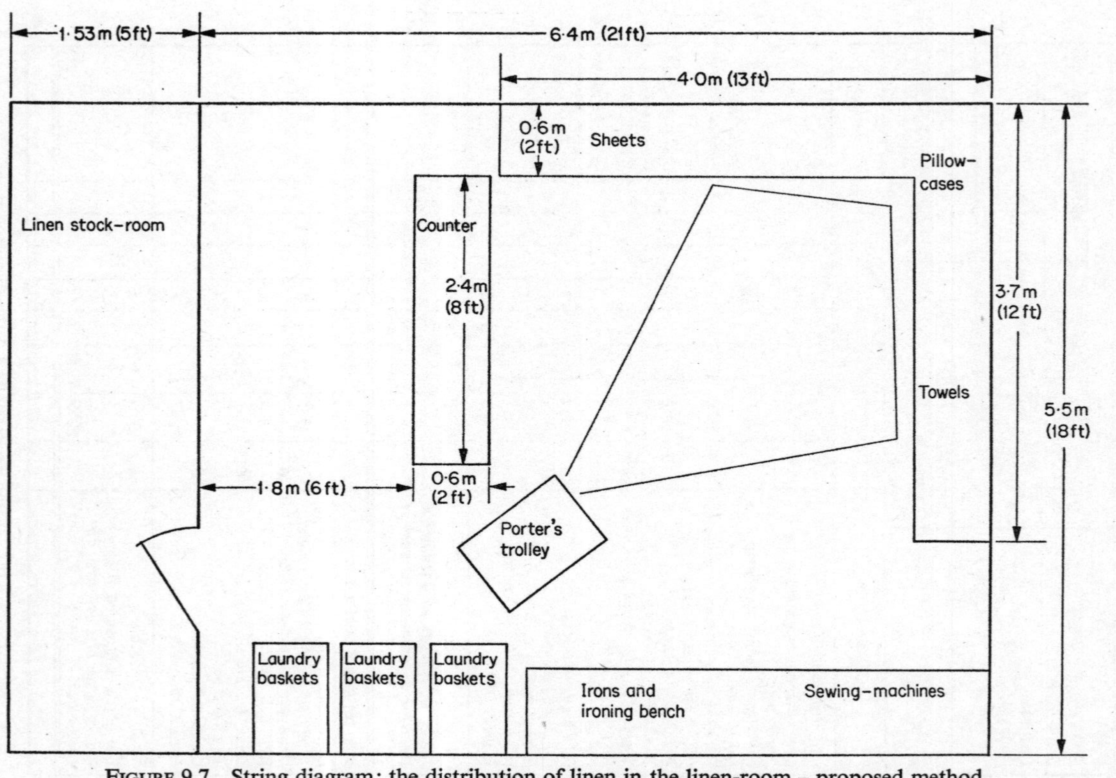

FIGURE 9.7 String diagram: the distribution of linen in the linen-room – proposed method
To issue 60 sheets, towels, and pillowcases

and bathmats. The weight the trolley would have to carry would be about 68 kg (150 lb).

It was decided to use a trolley 1016 mm long by 711 mm wide (40 × 28 in.) having three tiers or compartments; if this was found to be satisfactory, the housekeeper would then buy a battery-driven trolley, which could be usefully employed elsewhere in the hall when not needed for linen distribution.

ASSESSMENT OF STAFF REQUIREMENTS

A second example of the application of Work Study techniques is used when estimating the *staffing requirements* in new buildings or when re-assessing the staff establishment of an older concern. Staffing should be re-assessed periodically as working conditions can change radically over the years. New equipment is brought into use, newer and easier to maintain floorings and fabrics are used, and rooms may be adapted and used for different purposes. All this can alter the work load, often making an unfair distribution of work in the building, with the result that Nellie can now have quite an easy time whereas Mary is always busy and resents Nellie's ability to find time for a quiet gossip on the corridor. Moreover, the habit of someone 'helping out' at odd times can become accepted as a necessity, so that if she leaves a replacement is considered imperative.

Assessment of staffing requirements can be adequately carried out only when based on a sound knowledge of the 'average' normal time needed for any procedure with an allowance made for individual circumstances. Older buildings generally need more labour time owing to the vast variety of floorings and furnishings acquired piecemeal over the years. Moreover, the layout of the older building may not have been planned with the needs of the housekeeping department considered and arranged for; this does of course often apply to new buildings as well.

However well-known and familiar the building is, the basic procedure is to:

1. Obtain the plans of the building and study them. This will show floor, room, and stair lay-out, space available for cleaners' cupboards, power points and water supply, and the floor area with all rooms and floors numbered for easy reference. This study helps to eliminate all preconceived ideas on traffic lanes and area sections.
2. Using a form similar to the one shown in Figure 9.8, a physical survey of all rooms and corridors is made with an estimation of the time required for cleaning each area weekly.

This type of survey summarizes much useful information:

1. The total area.
2. Area and amount of each type of flooring to be dealt with.
3. An inventory of furniture, room by room.
4. It should show any special requirements or precautions which have to be taken, such as special cleaning for antique furniture,

Floor 2 *Block B*				Cleaners' stores *by fire escape*		Waste disposal	
Total area *353 m² (3 800 ft²)*				Power points *5 A each room 13 A 2 per corridor*		*Paper sacks*	
				Water supply *in cleaners' stores*		*Porters collection*	
Room No	Area, m²	Type of floor	Walls	Furniture	Special cleaning required	Remarks	Time assessment
18 bed/ sitting	*11·2 (120 ft²)*	*PVC + rug*	*Wall-paper washable*	*Divan built-in wardrobe and chest of drawers easy chair*	*None*	–	*1½ hours*
19	*13 (140 ft²)*	"	"	"	*Wash-basin*	–	*1¾ hours*
20	*20·5 (220 ft²)*	*Carpet*	"	*6 easy chairs 2 coffee tables sideboard mirror 2 pictures*	*None*	–	*2 hours*

FIGURE 9.8 Room survey – to show cleaning time assessments

office waste which must be treated as confidential and requires burning, or any restriction on cleaning times.

5. The total time required for cleaning each area. The time assessment will depend on the standard and frequency of the cleaning to be carried out. Some areas, such as public cloakrooms, reception areas, and public rooms, may require attention two or three times a day.

All times assessed are totalled. This will give the gross man- or woman-hours required for cleaning the building each week. This figure is then broken down into the numbers of full or part-time cleaners required. Whether full or part-time cleaners are employed depends on:

(*a*) The labour that is available.

(*b*) The times when cleaning can most conveniently be done.

Employing part-timers can involve extra costs due to:
National Health contributions (employer's contribution)
Administration costs
Uniform
Equipment
Staff facilities

A gross assessment for cleaning of 1150 hours on the basic 40-hour week would mean employing 28¾ full-time cleaners at 25p per hour at a net cost of £287·50 per week to which 18 to 20% is added to cover extra administration and insurance costs etc. In this way, the estimate or budget for labour costs can be arrived at on a yearly basis; and it is easy to calculate that if wages go up by 1p an hour then, to maintain the same standards and work rates, the estimate will have to be increased by approximately £500 a year. When times and numbers of cleaners have been agreed on, then the floor plan is divided into logical working areas and staff is allocated as necessary. Detailed work or instruction cards are written for all staff, and training is given so that the correct method of working and correct use of equipment are put into operation.

Instruction cards are of particular advantage as they lay down the standards and methods expected, and can be referred to in times of doubt and trouble. In an emergency or the staff shortage which occasionally occurs, a new cleaner, with the aid of her written instructions, should be able to make a reasonable attempt at the work on her own.

All initial planning must be reviewed after the first few weeks of operation so that difficulties and any objections can be analysed and put right.

In addition to the cleaners required, provision will have to be made for relief cleaners and for supervisors or forewomen. This depends on the numbers, holiday entitlement, the type of work, and the general sickness and absentee level expected.

A second method of calculating staffing requirements is to work on a fixed average time for the different types of accommodation as:

[*Example for Hospital Staff Accommodation*]	*per week*
For junior staff, with no service or bedmaking	1½ hours
For middle-grade staff, bedmaking but no other service	2 hours
For senior staff with bedmaking and service	2½ to 3 hours
Unoccupied rooms, kept ready for occupation	¼ hour
Toilets	20 min each
Bathrooms	30 to 40 ,, ,,

A third method is to calculate, as some of the Hospital Groups do, on floor areas:

Rooms or area under			per week
9·3 m^2	(100 ft^2)		1 hour
9·4 to 23·3 m^2	(101 to 250 ft^2)	–	2 hours
23·3 to 37·2 m^2	(251 to 400 ft^2)	–	3 hours
37·2 to 46·5 m^2	(401 to 500 ft^2)	–	4 hours
46·5 to 55·7 m^2	(501 to 600 ft^2)	–	5 hours
55·7 to 65·2 m^2	(601 to 700 ft^2)	–	6 hours

It follows that if an area of 9·3 m^2 (100 ft^2) requires 1 hour's attention each week, then 93·0 m^2 (1 000 ft^2) will mean 10 hours' work, and an area of 372·0 m^2 (4 000 ft^2) would require one cleaner working a 40-hour week.

Both these last two methods may seem arbitrary but are based on experience and averages of observed times. However they take no account of any of the special requirements or problems, such as age and state of the building or the type of occupant, which can so afflict the housekeeper. Time spent on planning is important as it enables one to look at the work objectively and means that there is less chance of making mistakes or wasting time and labour later on through haphazard methods of working. Tools and equipment can be checked and ordered in advance; there is time to explain to all concerned what plans or alterations will be needed and, most important, it is easier to know and communicate exactly what is needed, where and when.

10 GENERAL SAFETY PRECAUTIONS AND SECURITY

THE Offices, Shops and Railway Premises Act 1963 applies to all offices and shops (including catering establishments open to the public) and to canteens which cater wholly or mainly for persons employed in office, shop, or railway premises. Whilst the Act does not cover the residential part of hotels and other residential establishments, it does however apply to the office parts of these buildings, and it seems sensible to maintain the same standard of requirements throughout the establishment whether this is a legal requirement or not. Legislation may, indeed, be shortly introduced which will protect all employees in hotels and residential establishments.

RESPONSIBILITY FOR SAFETY

This is often shared between various departments: the contracts or finance department which arranges insurance cover and the provision of fire fighting equipment contracts; the estates or maintenance department which is mainly responsible for the lay-out and upkeep of the fabric of the building; and the housekeeping department which is responsible for its day-to-day running and which is in a position to act on and report any possible hazards as they occur.

FIRE PRECAUTIONS ACT

The Fire Precautions Act of 1971 is concerned with the protection of life in the event of fire and applies to all hotels, boarding houses or residential establishments providing sleeping accommodation for more than six persons, being guests or staff, and provided that this accommodation is above the first floor or below the ground floor.

To ensure that a building has reached the required safety standard a certificate will be issued by the local fire authority who have the right to inspect buildings and to recommend and enforce improvements.

The fire certificate specifies:

(*a*) the use of the premises which it covers;

(*b*) the means of escape in case of fire;

(*c*) the maintenance of the means of escape at all times – this would relate to emergency lighting, direction signs, and smoke-stop doors;

(*d*) the means of fighting fire for use by the persons in the building;

(*e*) the means for giving warning in case of fire.

The fire authorities may also require that members of staff should be trained to deal with fire emergencies and that records are kept of this training and may also limit the number of people that can be on the premises at any one time.

FURTHER RECOMMENDATIONS

The recommendations say that all staff should be given two periods of verbal instruction within one month and for new staff as soon as possible after appointment; the instruction should be reinforced by further half an hour periods every six months for day-time staff and every three months for those on night duty. In addition, all staff should be given a written copy of the Fire Instructions.

FIRE PREVENTION

The wording of the 1963 Act states that 'there must be appropriate fire fighting equipment, properly maintained and readily available for use'. Fires occur when any combustible material comes in contact with a source of ignition. The main sources are: matches and cigarettes which have not been stubbed out but left to smoulder; heating systems either overheating in the flues or producing sparks or hot ashes; electrical equipment and wiring which is either faulty, not earthed correctly, over-loaded, or with insulation which has deteriorated; equipment left 'on' so causing over-heating of fats or other food-stuffs in the kitchen; and direct ignition.

One cause which is fairly common is through the sun's rays striking a concave mirror – such as is used for shaving – which concentrates and directs the heat source on curtaining or other fabrics and will, in time, produce fire.

It is difficult to eliminate all possible causes of fire but risks can be recognized, and reduced by ensuring that all staff know the hazards and the right action to take and are trained to use the equipment provided.

Much can be done to prevent the spread of fire in a building, and advice is readily obtainable from the local Fire Prevention Officer. The main safeguards are the fitting of smoke-stop or fire-doors at key points in the building; these would be at access points to stair-ways or lift-shafts and between different floors and sections, so limiting the spread of fire. These doors must always be kept shut and never allowed to be wedged open by the staff. Some authorities stipulate the use of fabrics, curtains, and carpets which are treated with a fire-retardant, or the use of fabrics which will not burn easily.

Good housekeeping prevents the accumulation of rubbish, paper, and greasy rags in cupboards, under the stairs, or in rooms which are seldom used. All rubbish should be removed regularly and stored safely until disposal. Potential hazards such as open fires should be well guarded, and portable fires should be used in positions where they cannot over-heat curtains, and where cupboard doors cannot be opened against them. Ashtrays should be provided.

It should be part of an evening or night porter's routine to check the building after the staff have left and at regular intervals to make sure that no danger exists; this can be done at the same time that he checks for doors or windows left open and the general security of the building.

FIRE FIGHTING EQUIPMENT

CLASSIFICATION OF FIRES

Fires are classified as: Class A – which are those caused by wood, cloth, paper, or other similar combustible materials; Class B – which are caused by flammable liquids such as fat, oils, and petrol; and Class C – which are those caused by electrical equipment and wiring. A different type of extinguisher is required for each.

FOR CLASS A

The most suitable and effective method for this type is the use of water in an extinguisher; those of the water/CO_2 type produce a jet of water 9 to 10 metres (30 to 35 ft) long for about one minute. In large buildings fire hoses are installed which either turn on automatically when unrolled or are fitted with a wheel turn-on device. Soda-acid extinguishers are gradually going out of use.

FOR CLASS B

These should be extinguished by smothering. Water used on this type of fire will have the effect of spreading the flames quickly. Dry powder, foam, or carbon dioxide which smother the fire cutting it off from the air, are the types of extinguisher most usually installed. An asbestos or glass-fibre blanket has the same effect.

FOR CLASS C

These must be extinguished by smothering but with an agent which is also a non-conductor of electricity; foam or water will spread the fire and may lead to the danger of electrocution. Vaporizing liquids, such as B.C.F. (bromochlorodifluoromethane) which is replacing carbon tetrachloride extinguishers, are effective.

Extinguishers are normally hired from a company specializing in their manufacture and maintenance. These firms service them regularly.

FIRE DETECTION SYSTEMS

These are of two main types.

1. *Sprinkler installations* which are built into the construction of the building as a system of pipes incorporated in the ceiling and connected to the main water supply. Situated at regular intervals in the piping are outlets or sprinklers which are designed to open when the temperature is raised to a pre-determined level, often set to 74°C (165°F); for this reason they should never be cleaned with very hot water. When operated the water is sprayed over the area beneath each sprinkler and over-laps the area of its neighbours so that the entire room is covered.
2. On the same principle, *detector bells* are mounted in the ceiling and, if a fire breaks out, the sudden rise in temperature automatically rings the alarm bell.

FIRE ALARM SYSTEMS

All alarm systems are wired to a central indicator panel placed in a main office or hall where it can be easily seen; this quickly locates the source of the alarm so that appropriate action can be taken. Since it is only in very large or important establishments that an automatic alarm system has a direct link with the local fire brigade, instructions must be given to staff that, in the event of fire, the fire brigade have to be notified. Alarm systems usually have an emergency power source in case of a failure of the main electricity supply; this is by batteries which are kept charged either by a trickle charger from the mains or by a periodical charge. The alarm system should be tested regularly; the requirement of the Offices, Shops and Railway Premises Act being four times a year.

FIRE PRACTICES

These must be held regularly so that both staff, residents, and guests know what action to take and the most practical escape routes from the building. Fire escapes must be clearly marked and access to each must be kept unblocked and clear. Emergency lighting may be required for escape routes and to signlight the location of these doors. (*see* page 157 for security and type of door).

All staff must know what action to take in case of fire; this knowledge should be part of the induction training of all newcomers. Introduction to be given to staff includes:

1. Notify the fire brigade of the location and size of the fire, however small.
2. Fire fighting by the staff must only be undertaken if the fire is small and can be easily contained with no risk of loss of life to staff or guests. The local fire brigade in an urban area can usually arrive in 4 to 5 minutes of the alarm being given.
3. Doors and windows must be shut; power appliances should be turned off. Lifts must be switched off and must not be used.
4. When the building is cleared, all staff and residents should go straight to the assembly point and stay there until their presence clear of the building is checked and until it is declared safe to return. Warm clothing should be worn, but on no account should anybody go to cloakrooms or upper floors to obtain it.
5. The register of guests, residents, and staff should be removed from the building so that all can be accounted for.

Some onus must be placed on the *guests* that they become familiar with the action that they must take. Instructions placed prominently in each room might be as follows:

Fire

In the interests of safety against Fire, please take the following action:

On arrival. Find the Fire Alarm and Fire Exit nearest to this room.

If You Discover a Fire. Give the Alarm by breaking the glass in the nearest Fire Alarm.

If You Hear the Fire Bells Ring. Put on a coat and go by the nearest Fire Exit to the forecourt to the right of the main entrance.

Please close all doors behind you. On no account use the lift.

Management should know the time needed to clear a building and they should be confident that staff are trained so that action taken becomes a routine and will be carried out whether they are there or not.

ACCIDENT PREVENTION

The first legislation affecting the safety and welfare of employed people was introduced in 1802 in the Health and Morals of Apprentices Act. Since then there has been a series of legislation, the latest being the Factory Acts which embodied most of the previous rulings on safety, but only applying to those working in factories or on factory premises. In 1923, a Bill was sponsored by the Railway Clerks Association to give office workers the same protection; but this Bill was not passed and in 1936 and 1939 further efforts were made by the T.U.C. The Government set up a committee in 1949 but in spite of the report (the Gowers' Report) recommending legislation no Bill was introduced to the House until a private member's Bill was passed in 1960. This Act was superseded by the government act of 1963, the Offices, Shops and Railway Premises Act.

The Act applies to all people employed in offices and shops, to most railway premises, including catering establishments open to the public, and to canteens catering for all employees. Protection is still not given to employees in other types of establishment or work but, even if they are not covered by legislation, the 1963 Act is a guide to the conditions and minimum standards which it is sensible to maintain.

In addition to statutory law, all employers have a common law duty to take all reasonable precautions and care so as to prevent any likely injury or harm occurring to staff, residents, or guests. Failure to do so could be regarded as negligence and could lead to action for damages against the employer. Responsibility for the reasonable safety of all persons on the premises is not, however, a guarantee of safety under all conditions; and if an accident does occur which is considered to be caused by circumstances which it would not be reasonable to expect, or if the employee himself has been gravely at fault, the occupier or employer may not be liable, or blame may be allocated proportionally between employer and employed.

More people are killed and injured in the home than on the roads. According to the statistical reports of the Royal Society for the Prevention of Accidents (RoSPA) approximately one in seven or

15% of accidents occur in Residential Establishments; of these accidents, the greatest number are caused by *falls*. Falls on the same level, through slipping, tripping, or tumbling, are the most common, with falls on stairs a very poor second.

General Accident Prevention is closely connected with good housekeeping and involves:

1. The provision of non-slip floors and, if polish is used, non-slip polish.
2. The securing of mats and rugs.
3. Carpets should be tight and well-secured and repaired when worn. Stair carpet should be well-fitted and always in place.
4. Where one floor covering gives way to a different type, the join should be covered by a metal strip or the surfaces well jointed together. All floors should be kept in good repair.
5. Dark corridors, stairs, and steps should be lit possibly with strip lighting under the stair 'nosing' or inset near the floor. Lighting in these areas should always be left on, particularly in places where there are elderly or very young people.
6. Any hazards, such as low doorways or unexpected projections, should be well marked.
7. There should be sufficient space for free, easy movement within the building so that staff and residents are not constantly walking round or moving furniture.
8. Any open fire should have a substantial fixed guard. Gas and electric fires must have an adequate guard fitted to them before they are sold. Guards are a legal necessity for all fires whenever there are young children under 12 years of age, even when the fire is in a private home.
9. The use and care of equipment. Under the 1963 Act, some of the equipment in general use in residential establishments is listed as dangerous. These are: power-driven machines such as dough mixers; food-mixing machines when used with the attachments for mincing, slicing, chipping, crumbing, or any other cutting operation; pie and tart-making machines; vegetable slicers; whether power driven or hand-operated, all circular knife-slicing machines used for bacon and other foods; and potato chipping machines. The Act states 'No person under 18 years of age may clean any machinery if this exposes him to risk of injury from a moving part of that or any adjacent machinery'. It goes on to state that 'No person may work at any machine specified by the Ministry as dangerous unless he has been fully instructed as to the dangers and the precautions to be observed, and either has received sufficient training in work at the machine or is under adequate supervision by an experienced person'.

Some firms safeguard themselves in this respect by the use of an equipment book which is signed by all new employees after they have been instructed in the use and care of each piece of equipment.

10. All equipment should be well maintained. This includes the earthing of power appliances, fuses of the right amperage, and plugs intact and not broken and repaired by Sellotape as is occasionally seen.
11. A further section of the Act states that 'no person may be required, in the course of his work, to lift, carry or move a load so heavy as to be likely to cause him injury'. Figure 10.1 sets out instructions on how to lift.
12. Walking round any establishment quickly pin-points any likely causes or potential accident situations relevant to that building on which action should be taken.

FIGURE 10.1 Notes on How to Lift

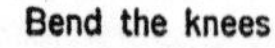

Bend the knees

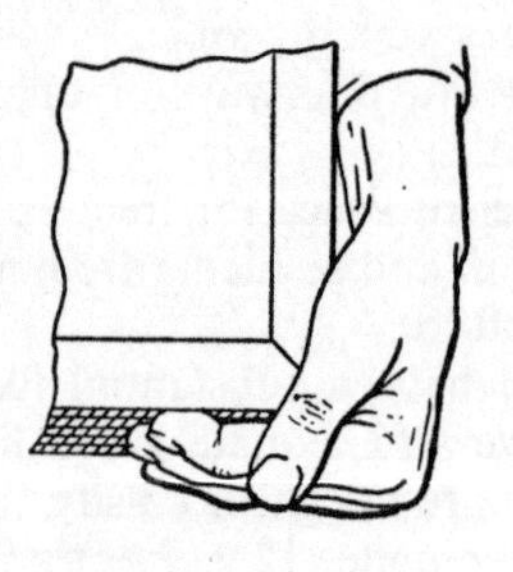

Proper grip

Grip at opposite corners
Arms close to the body

The secret of lifting and carrying without strain is to lift by bringing into use the strong muscles of the legs and thighs and not to use the comparatively weak muscles of the back.

Consider the size, weight, and ultimate destination of the load; *if the object is too heavy, get help.*

What to do	*Key points to mention*
Stand with feet apart	200 to 300 mm (8 to 12 in.) apart with one foot in advance in the direction in which one intends to go
Bend to a crouching position	Bend knees and keep back straight but not necessarily vertical. Keep chin in and avoid dropping head backwards or forward

[*contd.*

What to do	*Key points to mention*
Pick up with a firm grip	Use palm of hand; if finger-tips only are used, there is more strain on muscles and more chance of dropping the load. Grip opposite corners of the load. Keep arms close to body so that the body muscles will take the weight, not those of the arm, wrist, or shoulders
Lift	By straightening the legs, keeping the back straight. Lift load from floor to knee, then knee to carrying position
When carrying	Hold close to body. Do not attempt to change grip; when this is necessary, put load on a support and then change
To put load down	Reverse the lifting procedure

First Aid Boxes must be provided, and all staff must know where they are kept. As has been said elsewhere, the housekeeper and her staff are frequently called upon to give first aid and are well advised to seek one of the Red Cross or St. John's Ambulance First Aid certificates.

It is difficult to assess the cost of an accident as no reliable figures are available, but in 1968 21·9 million working days were lost through accidents at work. These were for reported accidents only and compare with 4·6 million working days lost through strikes in the same year.

When assessing the cost of an accident, the main factors to be considered are:

(*a*) The time lost by the injured person.
(*b*) The cost of the medical treatment.
(*c*) Interruption in the work schedule and the cost of replacement of the person injured.
(*d*) The legal costs.
(*e*) Not least, the effect on the person injured and on his family.

ACCIDENT BOOK

This book should be filled in as soon as is possible after an accident whilst it is fresh in the memory. Even if the accident seems to be trivial it should be recorded, as the extent of an injury is not always immediately apparent. As well as the more obvious accidents which have entitled the employee to injury benefit, other disabilities may be as follows: severe back-ache which developed several days after slipping on a wet floor, dermatitis resulting from the use of detergent, and a hand injury due to the too forceful wringing out of dish-cloths.

When recording an accident, the following information is required (*see* Figure 8.1): the date and time, the name and address of the person concerned, what happened, why, and where and any witnesses. This statement should be signed as correct by the injured person as soon as possible.

INDUSTRIAL INJURIES OR ACCIDENT

An Industrial Accident is one 'arising out of and in the course of' the injured person's employment. If, as a result of the accident, incapacity, disablement, or death occurs benefits are payable from the State Insurance scheme provided that contributions have been made (*see* Chapter 8).

If an employee is away from work for more than 3 days by reason of an industrial injury or if there is an accident which causes death, the local authority or H.M. Inspector of Factories (depending on which is the enforcing authority) must be informed.

SECURITY

The security of buildings can lead to many problems, particularly when the establishment is open to guests, staff, and residents over a 24-hour period. The complete safeguarding of property seems to be impossible but some action can be taken to mitigate loss.

1. Staff must be instructed to question anyone on the premises whom they consider to be acting suspiciously and whom they do not know.
2. There should be a routine inspection of the building by housekeeping or portering staff to ensure that doors, windows, stores, and offices are locked after duty hours. One method of making sure that this is done is by the use of a Night Porter's clock which registers the time of inspection when used with specially fitted keys kept in strategic parts of the building; this is only effective however when the clock is checked regularly by the office staff.
3. There should be control on all keys and master keys.
4. Where money has to be kept on the premises, it should never be obvious but should be kept locked away in a safe and not left about in open drawers or desks. The routine for handling cash should be as varied as possible.

5. Preventive measures should be taken to stop unauthorized access to the building. The most usual methods are by fitting key-operated locks or bolts to ground-floor casement or sash windows and the use of metal spikes or non-drying security paint (anti-climb paint) on all piping and guttering which leads to balconies and upper windows. Fanlights can be reinforced with a steel grille fitting.
6. Fire escape doors may cause difficulty as the means of opening must be always available; this means that they can also be used by staff and residents instead of the normal exits and can, in consequence, be left open. There are three main types of fastening:

(*a*) by panic-push bar. These can only be opened and closed from the inside.

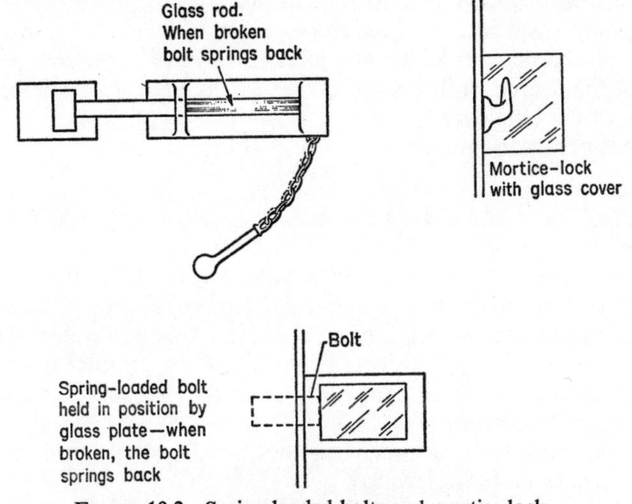

FIGURE 10.2 Spring-loaded bolts and mortice lock – available for fire-escape doors

(*b*) by lock and key. The key is kept, readily available, in a glass container by the door. This key should be tested occasionally to make sure that the lock works without difficulty. It has been known for the key to be removed and duplicated or replaced by another, so that an occupant can have free access to the building.

(*c*) by a spring-loaded bolt which opens as soon as the retaining glass is broken (*see* Figure 10.2).

Crime Prevention Officers from the local Police Authority are always prepared to advise on security when asked to do so.

PILFERING

Although the great majority of the staff employed are completely trustworthy, the occasion does sometimes arise when the Management or the Housekeeper has reason to suspect that pilfering is taking place. It must be stated that no private person has the right to search another person without their consent; to do so would be an assault on that person. Many firms therefore incorporate in the contract of employment between them and the employee a clause which does give the firm some rights to search. The wording might be as follows:

> The XXX Hotel reserves the right to stop and search all employees entering or leaving the hotel premises and to inspect the contents of their bags and parcels.

Before any search is undertaken, the Housekeeper should be reasonably suspicious that an employee is carrying stolen articles; the Housekeeper would be well advised to have a witness present during the search both on her own account and, if asked for, on behalf of the employee.

Other points to note are:

1. The right to search should be retained by the Management or Housekeeper and should not be delegated to other more junior staff.
2. Management should always be discreet and seek the co-operation of the employees stressing that no particular person has been singled out – the action is being taken in the best interests of both the employees and the firm. There should be no discrimination.
3. It need hardly be said that female employees should never be searched by a male.
4. A reluctance to be searched may not indicate guilt but may only be extreme embarrassment.

If a member of staff has been found with stolen property it should lead to dismissal; failure to dismiss will only mean that staff will feel that Management is lenient and it may encourage others to pilfer. Whether the employee is prosecuted will depend on the circumstances and judgement of the effect such adverse publicity may have on the firm.

REFERENCES

HMSO Publications

Student Residence; Building Bulletin 37 – University Building Notes (Department of Education and Science, 1967).

Training College Hostels; Building Bulletin 15 (Ministry of Education, 1957).

Space in the Home (metric edition); Design Bulletin 6 (Ministry of Housing and Local Government, reprinted 1970).

Residential Accommodation for Staff; Hospital Building Note 24 (Ministry of Health, reprinted 1967).

Flatlets for Old People (reprinted 1962).

More Flatlets for Old People (Ministry of Housing and Local Government, 1960).

Building Research Station

Hostel User Study, by Phyllis Allen (Current Papers, 1968).

Reports to the Council of Industrial Design

on *Carpets*, *Furniture*, and *Tableware*.

King Edward's Hospital Fund for London

Floor Maintenance in Hospital Wards; a progress report by F. E. Burnham.

Report on the *Washing-up of Crockery*.

National Economic Development Office (Hotel and Catering EDC)

Service in Hotels (1968).

Staff Turnover (1969).

Hotel Accounting (1969).

AJ Metric Handbook (The Architectural Press, revised edition 1969).

Introduction to Work Study (International Labour Office, revised edition 1969).

The Seven Point Plan; N.I.I.P. Paper No. 1.

Interviewing for Selection; N.I.I.P. Paper No. 3.

Employment Interviewing, by J. Munro Fraser (Macdonald & Evans).

Housekeeping Manual for Health Care Facilities (American Hospital Association, Chicago).

Employers' Guides; issued by the Government Departments on: P.A.Y.E., National Insurance, Graduated Pensions, Contracts of Employment Act, Redundancy Payments Act, etc. (Copies of Acts can be obtained through H.M.S.O.)

INDEX